Advance Praise

"All leaders must realize that it's their people, and not technology, that will be the biggest competitive advantage for their organizations to succeed in this new world. As human beings, we have the ingenuity, the talent, the intelligence, and the knowledge to create opportunity from all the disruption and change we are experiencing. I highly recommend this great book by David and Carlos. *The transHuman Code* should be on everyone's required reading list!"

—**Leena Nair,** chief human resources officer of Unilever, global champion for the establishment of human-centric leadership initiatives, including diversity and inclusion, for over 160,000 employees in 190 countries

"Human society is being transformed by new technologies, and the big question is: What is the DNA of this emerging new world? David and Carlos have assembled a tremendous resource for understanding what this transHuman world will look like and, more importantly, how we can guide it to be a world whose character is more human than technological. A must-read!"

—**Professor Alex "Sandy" Pentland,** co-creator of the MIT Media Lab, director MIT Connection Science, one of the most cited scientists in the world, a serial entrepreneur, and author of *Trust::Data* and *Social Physics*

"We are living in a time of profound disruption, which is creating both unparalleled opportunity and threat. It is essential, as we navigate our way toward our uncertain future, that the best of humanity is strengthened and protected and not irrevocably compromised. Strong, thoughtful leadership and creative ideas are needed. This incredibly important book provides just that."

—**Julia Christensen Hughes,** dean of the College of Business and Economics at the University of Guelph, advocate and activist for student-centered transformational leadership, and global innovator merging business and corporate social responsibility study

"The new digital age, or the new AI age, poses an interesting paradox for us. On one hand, it promises us the opportunity to live richer, fuller, and more rewarding lives by embracing the productivity and connectivity it offers. On the other, it poses a challenge of marginalization for some sections of the population and more broadly creates anxiety about the relevance of humanity. This book handles these questions deftly and comes down decisively on the side of the optimists, the dreamers, and the doers!"

—**Mohit Joshi,** president of Infosys and a global technology industry pioneer leading the company's innovative Banking, Financial Services, & Insurance and Healthcare and Life Science practice groups, which are transforming companies around the world

"In a time when climate change is making catastrophic weather events more frequent and scarcity of water more dramatic, we need to think how technology can be refocused to human needs. *The transHuman code* is reversing the machine-human balance to make our forward thinking start from the human element. It is a change of paradigm that we have the opportunity to realize with the help of the transHuman initiative. Everyone should read this book and contribute on finding the way."

—**Dr. Enrico Fucile,** chief of the Data Representation, Metadata and Monitoring Division, World Meteorlogical Organization, leading the gathering and predictive analysis of the increasingly important climate data from the 191 member countries

"With *The transHuman Code*, we can establish the most important basic principle of technological innovation—ethics of use. Every coder should owe a moral responsibility to only create products that the whole world can use safely in the world of AI. We have been waiting too long for this book!"

—**Kavita Gupta,** founding managing partner at ConsenSys Ventures, leading investment for the creator of the leading blockchain solution—Ethereum, a social finance pioneer and a leading innovator in technology financing and advancement

"This book is an essential reminder that human values require great attention no matter how impressive technology might become. The most powerful computer cannot replace human emotion, cannot substitute for ethical reflection, cannot render artistic expression obsolete. *The transHuman Code* forcefully urges us to embrace technology while keeping us—the readers, the humans—front and center."

—**Marc Firestone,** president of External Affairs of PMI, global advocate for innovative change, and co-leader one of the most dynamic corporate transformations in history

"We know where we are and where we are going to arrive; however, how we face that transition is a key element. David and Carlos are giving us a powerful platform where we can co-construct our best possible response as soon as possible. You have to read this book, and you have to interact and build with others the code with which we are going to produce our best way to evolve."

—**Elkin Echeverri Garcia,** planning and foresight director of the Ruta N Medellin, a serial entrepreneur, Latam technology investor, and change agent in the rise of "Smart Medellin" Columbia

"The future is, as always, uncertain. But now the range of possibilities is much broader than any time in human history. We can imagine the continuation of positive trends in the reduction of extreme poverty, decline in violence, enlargement of human rights, and expansion of democracy. It is obvious that technology accelerates everything. How we, together, respond and anticipate successfully is not at all obvious. David and Carlos have done us all an immense service by putting forward crucially important questions, accompanied by lots of data and information and enriched with provocative examples. All those who need to grapple with the challenges of the future, and that's all of us, will benefit from reading, pondering, and discussing this book."

—**Jack Faris,** chairman of The Global Alliance to Prevent Prematurity and Stillbirth, social communications pioneer, and multilateral social innovation champion

"Innovation is a creation by the minds, not from the code of machines. Therefore, we must work at the root of the minds to promote, accelerate, and deploy our best innovations: education of our young generations. David and Carlos' *The transHuman Code* offers a thorough study of how a well-balanced interplay of technology and education will be the key to determining our collective future and the path toward embracing the ever-so-important concept of 'innovation for good.'"

—**Marc Deschamps,** executive chairman of Drakestar Partners, one of the world's leading technology investment banking firms, who founded his first IT company at age 17 before leading major corporate technology initiatives across Europe

"In 1981, *The Soul of a New Machine* lauded the obsessive pursuit of a technology innovation that highlighted the feverish dedication of a few bright minds. *The transHuman Code* reminds us that humans remain at the center of automation's evolution, with a collective responsibility to bring our unique sensibilities to bear, whether we are technologists or not. *The transHuman Code* invites *everyone* to be a part of the conversation."

—**Beth Porter,** cofounder and CEO of Riff Learning, researcher and lecturer with the MIT Media Lab and Boston University Questrom School of Business, and an artificial intelligence (AI) pioneer

"This is a book that delivers a long-term view on how to manage the convergence of humanity and technology. Only David and Carlos have the foresight and network to bring together a stellar group of experts on the sociopolitical impact of techno-economical transformations happening on a daily basis all over the world. This is indeed a great platform to engage us all in a conversation that is so critical to our future!"

—**Danil Kerimi,** head of Technology Industries for the World Economic Forum, where he facilitates the critical global dialogue between government, business, and academic leaders on the future of technology

"Artificial intelligence, big data, robotics, sensors, and the Internet of Things promise a brave new world of greater efficiencies and novel solutions. What humanity needs, now more than ever, is a sensible user manual that will direct us in how to use this technology. Moreira and Fergusson give us this guidance. *The transHuman Code* is a lucid and compelling humanist's blueprint on how to ensure the disruptive technologies, which will define the 21st century, are harnessed for the public good."

—**Evan Fraser,** director of the Arrell Food Institute, Canada
Research Chair at the University of Guelph, one of the world's
foremost food authorities, and the author of *Empires of Food: Feast,
Famine, and the Rise and Fall of Civilizations*

"Our world is being transformed at an unprecedented pace. This book raises the question at the center of this new era of technological disruption: what will the role be for humans in the future? With *The transHuman Code*, we can now choose a path for our future that protects both humanity and human identity."

—**Don Tapscott,** executive chair of the Blockchain Research Institute,
author of *The Digital Economy* and *Blockchain Revolution*,
and one of the world's leading authorities on the impact of
technology on business and society

"Over the past 25 years, with China's rising, I have been witness to the most dramatic growth of the financial markets in our country. Today, China is undergoing another revolution, the technological revolution, and it is both promising and threatening. We are pioneering the development and application of blockchain, AI, and IoT for people of all ages, income levels, and professions. By applying *The transHuman Code*, China can bring together social responsibility and technological innovation for the benefit of all."

—**Wang Wei,** founder of the China M&A Group investment
bank and founding chairman of the China, Asia, and Asia
Pacific M&A Associations, often referred to as the "godfather of
Chinese M&A," is a global blockchain champion

"*The transHuman Code* is the must-read book of the year. As technology continues to disrupt every aspect of our lives, David and Carlos discuss the imminent need for a bold conversation on what makes us human and what values we need to preserve and strengthen—before it's too late. No topic is off-limits, and that's what makes this book so essential."

—**Megan Alexander,** correspondent for the #1 syndicated news TV show *Inside Edition* and author of *Faith in the Spotlight*

"Every user of technology—which is pretty much everybody—should read this book. It's filled with profound questions we should all be asking ourselves about what we hope our relationship with technology will, and will not, ultimately do for us. Before you pick up your phone again, read *The transHuman Code*."

—**Jon Rettinger,** co-founder of TechnoBuffalo, the largest independent consumer electronics portal in the world

"Whether humanity can maintain control of AI or not is a matter of the most profound importance for the future of mankind. Finding the appropriate equilibrium and the mental sanity over topics not related to a rational, logic, sequential discourse will be the greatest challenge of the 21st century. By opening the discussion under a platform of a civilized debate, Carlos and David are contributing to a holistic approach to the subject, one in which the brain, the soul, values, and feelings need to be properly incorporated and addressed. It is a healthy and opportune moment to have this discussion."

—**Rodrigio Arboleda,** chief executive officer of The Fast Track Institute, cofounder of One Laptop Per Child, architect, social innovator, and agent of exponential urban transformation

"We face a programmable world where the possibilities are limitless for humankind. The principles of the transHuman Code remind us that we, as creators and enablers of technology, have an obligation to ensure that our digital and physical environment will be programmed for the betterment of all. This is the handbook for the future we all deserve!"

—**Risto Siilasmaa,** Chairman of Nokia, cybersecurity pioneer, transformation engineer, technology investment angel, and author of *Transforming NOKIA: The Power of Paranoid Optimism to Lead Through Colossal Change*

Suzanne
Stay Real Fri.
Frankfurt !

THE
transHuman
Code

HOW TO PROGRAM YOUR FUTURE

CARLOS MOREIRA
& DAVID FERGUSSON

GREENLEAF
BOOK GROUP PRESS

Published by Greenleaf Book Group Press
Austin, Texas
www.gbgpress.com

Distributed by Greenleaf Book Group

For ordering information or special discounts for bulk purchases, please contact
Greenleaf Book Group at PO Box 91869, Austin, TX 78709, 512.891.6100.

Design and composition by Greenleaf Book Group and Kim Lance
Cover design by Greenleaf Book Group and Kim Lance

Cover image: koya79/Robot Hand Holding Planet/Getty Images;
Jolygon/Brain Illustration Isolated on BG/Thinkstock or as otherwise
shown on the Thinkstock website

Publisher's Cataloging-in-Publication data is available.

Print ISBN: 978-1-62634-629-1

eBook ISBN: 978-1-62634-630-7

Part of the Tree Neutral® program, which offsets the number of trees consumed in
the production and printing of this book by taking proactive steps, such as planting
trees in direct proportion to the number of trees used: www.treeneutral.com

TreeNeutral

Printed in the United States of America on acid-free paper

19 20 21 22 23 24 25   10 9 8 7 6 5 4 3 2 1

First Edition

CONTENTS

Introduction

HOW TO READ THIS BOOK

We know: The last thing the world needs is another book on technology. Let us set the record straight from the outset. While this book gives plenty of space in its pages to introducing you to some of the most potent technologies today, emerging from the most important industries on the planet, this is not a book about technology. It is a book about humanity—the role humanity is playing now and, most importantly, the increasing role it must play to ensure that our common values, the values that make life worth living, remain within our control. Yes, technology is here to stay. It is no exaggeration to assert that it's one of the most potent tangible forces, if not the most potent tangible force, that can improve our lives. But that's only if we are wise in how we use it, program it, and partner with it.

It is heavy-handed to say that technology is the devil or that it will eventually doom us. *Can* technology doom us? It's possible. *Will* it? Only if a few billion humans let it happen. This assumes, of course, that we're paying attention to what certain technologies are doing (and will do) to us. It's critical to our best future: discernment, vigilance. We don't need to be uptight about it, but we ought to readily acknowledge that sometimes, what is exciting and convenient in the beginning ends up being a wolf in sheep's clothing. We create technology as a solution—but sometimes, we end up creating a virus instead.

While the human world doesn't always make the best decisions for itself, and while we can easily overlook the forest for the trees where technology is concerned, we still learn, we still adapt, and we still work together to make things right. History has shown this much: We're fallible, but we're also unfailingly ambitious. Despite our mistakes—even the biggest ones—we continue to fight for a higher quality of life. We believe this innate drive enables us to always self-correct. This book hinges on the power of that truth, but we need to get busy applying it. If we're not wise about how we proceed in this season of our partnership with technology, it is more likely than most realize that our potential—all that is possible for us—will be scaled back to a level we've never experienced before. As interconnected and interdependent as we are on technology, a series of wrong decisions, wrong adoptions, and wrong consumptions over the next couple of years could lead to a decades-long regression—one in which human ingenuity is supplanted and human sensibilities are replaced by technology. We don't know how we would handle this, because we've never been threatened on this sort of scale before. What would we do if we could not be who we are? In the end, the question of the day isn't "Will we survive?" as much as it is "Will we still thrive?" The answer is still, collectively, up to us.

From here on out, it's best to consider this a book of conversations— the most important conversations in which we-the-world should be engaged as we aim to use the power of technology to script the best future possible. This book is not meant to provide solutions. How can it when so much of what is needed requires collaboration and adoption on a grand scale? We are merely two informed authors who aim to jump-start the innovation, collaboration, and adoption necessary to produce the solutions we ultimately (and, in some cases, desperately) need.

As a starting point, we'd like to offer a center-point on which humanity can already agree where technology is concerned: the following seven declarations comprise what can be considered humanity's critical technology proclamation—the transHuman Code Manifesto:

1. **Privacy:** Securing the privacy of every human being is paramount to realizing the full potential of our future. Therefore, personal data conveyed over the Internet or stored in devices connected to the Internet is owned and solely governed by the individual.

2. **Consent:** Respecting the authority and autonomy of every human being is paramount to realizing the full potential of our future. Therefore, personal digital data will not be used as research, rationale, enticement, or commodity by any entity or individual, except with the explicit, well-informed, revocable consent of the individual owner of the data.

3. **Identity:** Valuing the identity of every human being is paramount to realizing the full potential of our future. Therefore, everyone everywhere has the right to be known and validated by the possession of a government-issued digital identity, which can be authenticated and used only by its owner.

4. **Ability:** Advancing human faculties is paramount to realizing the full potential of our future. Therefore, to that end, the secure, approved, and accountable aggregation of personal information and resources to increase our individual abilities is a fundamental objective of technology.

5. **Ethics:** Improving the human condition is paramount to realizing the full potential of our future. Therefore, a universal code of ethics reflecting the highest order of human values will govern the development, implementation, and use of technology.

6. **Good:** Advocating and innovating the greatest good for all humanity is paramount to realizing the full potential of our future. Therefore, technology, no matter how advanced, will never supersede the spiritual purposes or the moral rights and responsibilities of any human being anywhere.

7. **Democracy:** Democratizing human vision, ingenuity, and education is paramount to realizing the full potential of our future. Therefore, technology will remain humanity's greatest collaborator but never represent humanity itself.

The solutions that humanity ultimately pursues must be developed and implemented collaboratively. If there is not global consensus about the inherent human values we aim to protect and enhance, and if we do not unify around the most critical decisions concerning their protection and promotion, human progress will be too fractured to move us forward. Finding the solutions we need, if we find them at all, will take longer than our lifetimes.

On the other hand, if we can agree on humanity's essentials—at least a majority of us—history suggests we will seek out and find our best answers as though they are necessary for our survival. In a way, they are.

You might be wondering, *How should we answer the critical questions of humanity where our future with technology is concerned?* What we propose is found in the title of this book: a transHuman code. Before we explain what this means, we'd be remiss not to first explain what that *doesn't* mean. Let us clarify any confusion by saying that this is not a book that espouses the movement known as transHumanism, which is framed by a belief that humans are not a fully developed species and, therefore, ought to be replaced over time by a more advanced bionic species—what some call humanoids. If you've seen the HBO series *Westworld* or the Swedish sci-fi drama *Real Humans*, such shows explore an earth in which transHumanism has been embraced to the point that a volcanic tension exists between humans and a human-like technological species. This future is neither one we believe in nor one we think is in humanity's best interest.

While some believe artificial intelligence is the god of the future, we believe that if this is true, AI is a god we must humanize; we are still creating a world for us, not technology, to thrive. Therefore, when we speak of a "transHuman" future in this book, and, in particular, when we use the phrase "transHuman code," we do so only as a nod to the reality that humanity as we know it is being *transformed* by technology. From our viewpoint, transHuman (with a capital "H," to remind us of the priority) is a term that merely points to our belief that our best future will come from a transformational relationship with technology—not one that requires us to surrender to a "better," more bionic species. In fact, we believe a *Westworld*-type future would be more bleak and gloomy than we can imagine. There are few circumstances more inhuman than that.

To avoid that fate, and to usher in the brightest, most fully human future we can imagine, we must commit to implanting—or coding—technology with the human values and attributes that promote and protect the human species as it is today. Said another way, we must develop a multifaceted, multi-industry strategy for programming human essence into the artificial intelligence (AI) we create, embrace, and consume.

This is not to say that AI is not useful. Using AI wisely is, in fact, the most profound opportunity we have to not only embed what is human into all future progress; it is also an ideal resource for determining what is most critically human in the first place.

Humans already have the greatest "code" on the planet programmed into us. "Simple truths," wrote the French essayist Vauvenargues, "are a relief from grand speculations."[1] This is a book about the simple truth that *humans* are the greatest technology on the planet. If enough of us embrace this truth, we have a real shot at ending global plagues like cancer, hunger, and AIDS. Innovation will flourish. Democracy will lead. Compassion will reach further.

If you doubt this truth, consider that your body is made of about 7 octillion atoms (or 7,000,000,000,000,000,000,000,000,000,000 atoms, if you're counting zeros). You have 37 trillion cells in your body, 50,000 of which will die and be replaced in the time it takes you to read this sentence. If your DNA were uncoiled, it would stretch 10 billion miles—or the distance from Earth to Pluto and back. In one day, your heart pumps 100,000 times, producing enough energy to drive a semitruck 18 miles. In a year, the amount of blood it pumps through your body amounts to approximately 3 million liters; the annual energy it produces can drive that same semi to the moon and back. Were your arteries, veins, and capillaries laid out end to end, they would extend over 62,000 miles— or long enough to reach around the Earth 2½ times. Your eyes can distinguish up to 7.5 million different colors. Your nose can differentiate between 1 trillion different scents. And then there's your control center: the human brain, powered by a trillion nerves, functioning in such a complex, and yet harmonic, manner that today's greatest neuroscientists estimate we understand only 5 percent of how it works.[2]

Perhaps we understand even less about how our DNA works. "Something inexplicable happens beneath the surface to form the all-knowing intelligence of DNA," explains Deepak Chopra in his fascinating bestseller *Quantum Healing*. "Sitting by itself in the middle of every cell, completely offstage, DNA manages to choreograph all that happens onstage. How does DNA manage to be the question, the answer, and the silent observer of the whole process at the same time?"[3]

"Without a doubt," wrote Dr. Werner Gitt, "the most complex information-processing system in existence is the human body. If we take all human information processes together, i.e., conscious ones and unconscious ones, this involves the processing of 10^{24} bits daily. This astronomically high figure is one million times greater than the total human knowledge of 10^{18} bits stored in all the world's libraries."[4]

And all that's merely the science of the matter.

We're not even scratching the surface of the artistic and spiritual complexities found in every human on the planet, or the abilities to love, to long, to dream, and to resolve. The human is and will always be the greatest and most advanced technology the world has ever known. Doesn't it then make the most sense to place the understanding, improvement, and utilization of humanity as today's highest priority?

This is our goal in the pages to come.

By introducing you to some of the most important developments occurring today, giving you a clearer understanding of their implications and then sparking the conversations that need to happen as a result, our hope is that, together, we will develop a transHuman code that will allow us to remain both the apex and axis of all technological progress from here forward. Thus far, human transformation via technology has been largely inspirational and passive, merely wishful thinking or hopeful vision where the majority of the planet is concerned. Now that we've seen and felt the volatile nature of our marriage with technology, we must get more involved and begin governing our future so that we are not governed by our own creations.

For the purposes of contextualizing the topics covered in the chapters to come, we suggest using Abraham Maslow's hierarchy of human needs (shown on the following page) as a general framework to help prioritize our conversations—and thus, our efforts.

Maslow's Hierarchy of Needs

This is merely a conceptual platform that can guide us to create, implement, and adopt technologies according to human priorities as we continue the conversations brought to bear in the book. While we will not provide an argument for the prioritization of technologies within the topics we will discuss, we do believe in the critical importance of a collaborative discussion surrounding this activity, and we hope that in highlighting a framework like Maslow's hierarchy of human needs, the positioning of our priorities will remain at the forefront of our minds moving forward.

If we are able to program a proper transHuman code into all tech efforts for the topics covered in this book, we will go a long way toward ensuring that humanity remains at the center of gravity in the universe—that our thriving will continue all the more. For each topic we cover in the book, therefore, we will describe the current state of things and then frame a discussion around the most important considerations (where our strategies and applications are concerned), pointing to key developments already happening (productive ones, unproductive ones, and counterproductive ones), and posing critical steps that ought to be taken in order to ensure human needs are met and humanity is advanced. To offer an ongoing context for these critical steps, we will use a shaded pyramid icon to indicate how productive efforts will specifically serve humanity's needs according to Maslow's hierarchy. We will also provide a brief recap of the most pressing questions that require answers today. Use these. Tweet about them. Write articles about them with your suggested solutions. Talk about them around your dinner table and boardroom. Post them to discussion boards, or head to ours (www.transHumancode.com), where specific and open discussions are already occurring on every topic we cover in the book—as well as many others.

We'd be remiss not to also recommend that these conversations extend beyond the individual to reach major decision-making bodies throughout the world: governments, corporations, nonprofits, NGOs, and so on. Our longtime work with the United Nations and the World Economic Forum has taught us that while changes, innovations, and improvements start at an individual level, they can quickly be scaled if major bodies are involved. If your work provides you access to such bodies, or influence within them, we encourage you to take the critical conversations that arise in and from this book to them.

Finally, a note about the functionality of the book you're holding. You are invited to engage directly in the development of the transHuman code. To date, the code for the major technological platforms that we all use—Facebook, Google, Twitter, Apple, etc.—has been developed by a small number of people. The transHuman code will be starkly different. It will be developed by humanity, all of us, potentially billions of people in conversation and collaboration with one another. It's an inclusive code—our code. It does not matter if you know how to program or, in fact, whether you know anything about how any technology works at all. You can play an important role in determining what sort of technology we not only want but need. This is another way of saying that if you're human, you ought to be involved in the decisions that surround what we will create and consume and why we will do so. This must be a global collaboration of people of all vocations, income levels, and ages living across the world. If technology impacts your life, you have an important voice. What we can build from here is essentially the GitHub or Wikipedia for the entirety of technological creation. It is a wholly open-source process, a wise, aggregate, collaborative codifying of our partnership with technology, now and into the future.

The opportunities that arise from the pages of this book fall into two categories: to contribute to the creation of the transHuman code (1) practically (doing the work of writing or implementing the code) and (2) philosophically (determining what should be protected and promoted by the code that is written into the technology you use). We hope that you discover—in the coming pages, if you have not already—that participation is the only option.

1

THE PINNACLE AND PURPOSE
OF TECHNOLOGY

Unfortunately, it's easy to lose sight of our preeminence in the grand ecosystem—especially during an era in which it is tempting to lean on technology to lead us into the future we desire. Do we really believe that technology—technology that we created, mind you—can become more complex and necessary than we are? Can the created ever really supersede its creator?

It's a question you have to answer for yourself. We all do. And together, we must collectively decide if we are building a better future for humanity with the help of magnificent technology . . . or building a future of better technology at the expense of humanity. There's really no simpler way to put it. The future is still in our hands. But a future is possible in which we are not in control. There would be no one to blame but ourselves.

We've actually been down a similar road before, and it didn't turn out well. We collectively chose wrong—or, better said, we didn't choose right soon enough. We didn't put humanity first, elevating the promises of technology instead. We let technology lead us, and it led us astray. This seemingly subtle oversight altered our lives for the worse, forever.

The year was 1895, and a German mechanical engineer named Wilhelm Roentgen discovered X-rays. His discovery was the precursor to French chemists Pierre and Marie Curie discovering radioactivity three years later. In the decade following the Curies' discovery, New Zealand physicist Ernest Rutherford and English radiochemist Frederick Soddy discovered that the radioactivity in uranium was the result of atoms splitting. These three discoveries alone—X-rays, radioactivity, and splitting atoms—catapulted the role of technology forward throughout the world, which in turn changed the course of human history for good and, because we weren't vigilant, for bad.

Roentgen won a Nobel Prize, as did the Curies, Rutherford, and Soddy, and Soddy's work on radioactivity brought nuclear reactions to the attention of the world, becoming the primary inspiration for H. G. Wells's 1914 futuristic novel, *The World Set Free*, which features atomic bombs dropped from planes during wartime. Twenty years later, the atomic bombs Wells imagined were becoming a reality in a secret lab in Germany. Hungarian physicist Leo Szilard penned a letter to U.S. President Franklin D. Roosevelt, apprising him of Germany's plans and urging the U.S. to begin developing its own nuclear weapons. Szilard enlisted Albert Einstein to sign the letter as his own, to give it as much weight as possible. The Manhattan Project was born.

Six years later, on August 6 and 9, 1945, the U.S. dropped their secret weapons on Hiroshima and Nagasaki, effectively ending World War II. The human loss was stupendous, but, in truth, the lasting damage to humanity had begun well before the summer of 1945—precisely 50 years before, when we allowed a string of scientific discoveries to elevate technology above humanity.

In our imprudent excitement to race into the future, our predecessors rode the wave of technology all the way to an irreversible threat called nuclear war. This realization prompted Einstein to eventually lament, "It has become appallingly obvious that our technology has exceeded our humanity."

The plot looks different today, but the stakes are no less momentous than they were when X-rays and atom splitting were discovered. Humanity is, once again, faced with choosing to play either the protagonist or antagonist in its own story—a story for which the conclusion has not yet been written.

The wave of advances preceding our current technological crest is just as noteworthy as the one during the turn of the 20th century— from Robert Metcalfe's first Ethernet (1973) and Cerf and Kahn's first Internet (1975) to Jobs and Wozniak's first personal computer (1976) and Berners-Lee's World Wide Web (1990). These thrilling advancements spawned the first browser, the first search engine, the first social network, the first smartphone, and the first app. Technology is now accelerating further, with developments like virtual reality, blockchain, digital currency, artificial intelligence, and robots.

Once again, we're at a tipping point, with a decision to make before atomic damage is done. We cannot afford to be naïve again. To make the right decisions—at least at the outset—we must begin by reminding ourselves that these bodies, minds, and souls we each possess are more advanced than anything in the world. It is this very resource—the human being—that provides us with our greatest inspiration and our greatest interpretation of the best future for all.

We must remember that all technologies that currently exist, and all technologies that will exist, are products of the consummate technological system within us. If we will embrace our principal value in this vast universe, and then harvest humanity's vast resources, we can ensure the days ahead will fulfill the deepest hopes of the greatest number of people. If we lose sight of this, we will lose more than lives: We will lose our reasons to live.

WIRED founder Kevin Kelly calls technology "humanity's accelerant."[1] This acceleration duly excites us. We celebrate together when Apple releases its latest iPhone. We feast on news of AI robots, virtual reality, and self-driving cars. Tech is the hippest topic and hottest investment segment in the world right now, and that doesn't look to slow down anytime soon. Many scenarios that once existed only in science fiction are reality today. This is both fascinating and promising. We can see a future of unprecedented efficiencies in industries such as education, health care, and electricity, as well as bigger victories—perhaps even lasting ones—over global afflictions like cancer and hunger.

But there remains a current conflict, as Erik Weiner points out in an op-ed piece for the *LA Times*. "We live in the Age of Convenience," he writes. "That concept lies at the heart of what Silicon Valley is selling and we are so eagerly buying. We see convenience not only as a nicety but an expectation, an entitlement . . . but too often we fail to recognize the full cost of our convenient lives."[2]

Today's technological world was built, and is governed, by the minds and resources of a few hundred thousand people. As a result, we are living in a society in which the richest 1 percent—most of whom made their fortunes in technology—have now accumulated more wealth than the rest of the world put together. Through our concessions to technology, we have unknowingly authorized an economy for the 1 percent instead of creating an economy that works for the prosperity of all, for future generations, and for the planet. This imbalance will be accelerated if we don't collectively remember the value inherent in humanity at large and begin to put the rights of people ahead of the rush to profits.

We-the-global-community have the opportunity to tap the minds and resources of 7.5 billion people, soon to be interconnected by more than 50 billion devices (by 2020) through the rapid growth of the Internet of Things (IoT)—the aggregation of all connected devices around the world. Consider the implications of this. The most efficient supercomputer in the world right now—a multibillion-dollar technological marvel from IBM and the U.S. Department of Energy's Oak Ridge National Laboratory (ORNL) called Summit—runs on a total of 200,000 CPUs (central processing units, i.e., computer brains) capable of processing 200 petaflops, a computing term that amounts to approximately 200 quadrillion (or 1 followed by 15 zeros) calculations per second. By comparison, IBM researchers have estimated that a single human brain can process 36.8 petaflops per second, or approximately one-sixth the computing power of Summit.[3] Said another way, six humans sharing resources equals one Summit. Consider what this means for humanity's ability to create, innovate, and problem-solve in a swift manner on a massive scale. If we can use technology to access the processing capability of the entire population—the original vision of World Wide Web creator Sir Tim Berners-Lee—we can ignite the equivalent power of more than a billion Summit supercomputers. It's no stretch to assert that the best future we can imagine for the most people is more available than we think. We just have to seek it more than short-term convenience.

The solutions and improvements we desire are within our grasp—many of them within our lifetimes, if we take the necessary steps to cement ourselves in the seat of transHuman authority and accountability, one technological advancement at a time, including those already developed and those still to come.

"At the center of every significant change in our lives today is a technology of some sort," writes Kelly.[4] The foremost matter that will determine what the world becomes is how we will use technology—not what we will use. Technology is here to stay, as a primary catalyst and dominating force. It is becoming, and will continue to become, a greater and greater part of our lives. This is very exciting news if we know that our place in the universal ecosystem is at the apex. The critical question is whether we will continue to allow technology to erode our prominence or reassert ourselves as the authors and perfectors of technology, whose job is to ensure that what is created, embraced, and proliferated always produces, first and foremost, better, healthier humans.

"We are different from our animal ancestors," explains Kelly, "in that we are not content to merely survive . . . This discontent is the trigger for our ingenuity and growth."[5] The decision we must make is to be wise stewards of our supreme ingenuity and growth. We must constantly ask ourselves, "What is prevailing: humanity or technology?" And we must do what is necessary to ensure our answer is humanity, and always humanity, the world over.

Some believe we should unconditionally render control of our future to the machines. They base their beliefs on something called the technological singularity, which hypothesizes that the artificial intelligence already present will eventually cause an intellectual explosion, resulting in a powerful computer superintelligence that would, qualitatively, far surpass all human capabilities. Sci-fi author Vernor Vinge says in his essay "The Coming Technological Singularity" that this will signal the end of the human era, as the new superintelligence would continue to upgrade itself and advance at an incomprehensible rate.[6] We would, in other words, become subservient to the machines.

The great fault in this hypothesis is that it does not account for the spiritual and moral characteristics of humanity, which set us apart from every species on the planet—characteristics like intuition, empathy, vision, conviction, and, to Kelly's point, ingenuity, stemming from a constant desire for better. There is a critical reason why Elon Musk recently confessed that humans are underrated.[7] A robot will never know how it feels to suffer from cancer or lose a child to starvation. Artificial intelligence will never comprehend the magnificence of childbirth, an ocean sunset, or the fulfillment of a decade-long dream. Computer programs will never match human complexity, with its wide range of emotions and tribal characteristics.

The best technology can do is prioritize its efforts according to *our* administration, *our* design, *our* programming. The key is making sure the priorities we assign to it and the governance we ascribe to it are in the best interest of all humanity. How? In general terms, the solution is to codify the core human attributes, those that set us apart from every life-form on the planet, into the technology we create; to write HI code into AI technology, so that the product serves us instead of making us subservient to it.

Technology only has the freedom to go as far as we allow it. To this point, we've been lethargic in managing its freedom. The manipulation of the U.S. presidential election was perhaps a turning point. In the least, it has served as a global wake-up call. But the truth we must grasp is much smaller: We're all complicit in the Facebook platform's capacity to manipulate. We created it and have embraced and empowered it for years, both knowingly and unknowingly.

In her article for *Vanity Fair*, Susan Fowler describes the conflicting human efforts that have allowed technology to initiate a swelling, covert coup against humanity. In her first weeks on the job as a programmer at Uber, a coworker, in hushed tones, encouraged her to remember the drivers when writing code. Fowler admits she didn't quite understand the meaning of this interaction until several weeks later, when she overheard two coworkers discussing ways to manipulate driver bonuses so that they could be fooled into working longer hours. "Shortly thereafter," writes Fowler, "a wave of price cuts hit drivers in the Bay Area. When I talked to the drivers, they described how Uber kept fares in a perfectly engineered sweet spot: just high enough for them to justify driving, but just low enough that not much more than their gas and maintenance expenses were covered."[8]

For a week in January 2012, Facebook removed between 10 and 90 percent of the positive emotional content from the newsfeeds of approximately 700,000 users. The action was part of a covert study being conducted by Facebook and U.S. academics, who were both interested in whether the emotions expressed by friends via social networks influenced users' moods. The ultimate objective was to determine if it was possible to manipulate users' feeds to keep them happier, which would, theoretically, keep them on Facebook longer, exposing them to more ads and increasing Facebook's revenue. According to *The Guardian* columnist Stuart Jeffries, the study found that "reducing the number of emotionally positive posts in someone's newsfeed produced a statistically significant fall in the number of positive words they used in their own status updates and a slight increase in the number of negative words."[9] In other words, the study proved that Facebook can, if it so chooses, sway users' moods to benefit the bottom line.

As you might imagine, when the news was leaked that Facebook ran the test, the response from users was less than cheerful. Explains Jeffries, "There was a disgust at the possibility that Facebook wanted to make us happier on their site so that we'd stay there longer . . . so that Mark Zuckerberg can buy more yachts." While Jeffries deems it "a loathsome business model," he acknowledges that the practice is nothing new in the world of technology platforms. He cites Thomas Jones from the *London Review of Books*, who reminds us that "the purpose of Facebook is to harvest, organize, and store as much personal information as possible to be flogged, ready-sifted and stratified, to advertisers . . . We aren't Facebook users, we're its product."[10]

This "mediation and commodification of every aspect of everyday life," as Jones puts it, is not the practice of Facebook alone. It is the modus operandi of nearly every technological tool you use, from cell phones and search engines to doorbells, security cameras, and voice-controlled speakers. The unveiled reality of the technologically driven world we live in can lead one to ask, as Jeffries does, whether we've become pawns "whose moods can be altered like lobotomised lab rats" to boost corporate revenues.[11] The truth is that the handful of multibillion-dollar platform companies who are vying for control of the web—Facebook, Amazon, Google, and Twitter, to begin with—aren't looking to influence our moods for the sport of it. These companies are, in most cases, looking to boost our morale while using their product and then feed us more of what we are looking for. It is, on one hand, deploying the fundamentals of good customer service. What better way to serve a customer than to understand the person as much as possible? However, the methodology gives rise to more critical questions:

- What if the information being compiled on you doesn't tell the full story?
- What if what you are looking for, aiming for, is deeper than a dozen digital imprints a day?

- Most critically, what *else* can be done with the information these companies have on you?

In a world where we are increasingly influenced by the technology we use, the more we use it, the more we lose our freedom to be human.

The pot of gold for modern technology companies is compiling, translating, and selling your identity, personal data, and behaviors to marketers of other companies, who need this information to sell you their products. This algorithm, called behavioral targeting, effectively uses technology's translation of your behavior to influence your future decisions. It sounds harmless at face value; it seems little more than an astute marketing strategy in the modern age. The easy conclusion is that we don't have to let it affect us—and perhaps you don't believe that it does. Unfortunately, this prevalent path for monetizing technology does more than improve corporate marketing efforts: It changes what we think about ourselves, which directly influences how we act and who we become.

A 2016 *Harvard Business Review* study showed that behaviorally targeted advertisements imply that we carry certain social labels, which we embrace because we believe our technology's conclusions are accurate— perhaps even more so than our own.[12] In the study, 188 undergraduate students were exposed to an ad for a high-end watch, which they believed was either targeted to them or not targeted at all. The test administrators then asked the students to rate how sophisticated they perceived themselves to be (the subjects had also been asked the same question before the test). The results show that "participants evaluated themselves as more sophisticated after receiving an ad that they thought was individually targeted to them, compared to when they thought the ad was not targeted. In other words, participants saw the targeted ad as reflective of their own characteristics. They accepted this information, saw themselves as more sophisticated consumers, and this shift in how they saw themselves increased their interest in the sophisticated product."[13]

HBR took these results a step further. They administered another study to determine if the changes in self-perception from the ads would extend to behaviors beyond purchases. In short, they did. This time, a group received a behaviorally targeted ad for an environmentally friendly product and, like before, subsequently rated themselves as more "green" than they had before the study. They were then asked to donate to a pro-environmental charity. Most were more willing to give money after receiving the targeted ad than before receiving it. In other words, the targeted ad telling them they were green swayed them to act more greenly.[14]

While this is a small sampling, it demonstrates that the permissions we've given to technology are not simply harmless, and the outcomes are not benign. Today's technology can mold who we are, what we do, and who we become—for commercial purposes, not humanitarian ones. If we are to become a better world, we each must flip that script.

While you might not mind the timely product and service suggestions dotting your inbox and flanking your screens, the implication of your daily concessions to technology is much more than an acceptance of commercial governance for greater convenience. At the heart of it all, behind a veiled reality few can see, you are outsourcing your humanity to a short list of companies who, while possibly well-meaning, can never fully protect you, never wholly represent you, and never facilitate the realization of your hopes and dreams. In most cases, they are doing precisely the opposite: undermining your basic autonomy.

This is much bigger than an economic concern. We're talking about a clear and present danger to humanity's livelihood in the universal ecosystem. We are the pinnacle of existence, the crown of creation. For this to remain true is no longer a forgone conclusion. We've unknowingly created our greatest nemesis in the global story—our modern-day Frankenstein. But we still have control over the story's outcome. We must wield that control willingly and wisely.

When Sir Tim launched the World Wide Web nearly 30 years ago, his purpose was not monetary. His vision was for it to foster collaboration between universities and scientists in an open, uncontrolled, and accessible manner. It quickly grew into a tool that expanded all of humanity's ability to learn from one another, help one another, and collaborate to improve the world. This was a beautiful thing. Today, however, Berners-Lee's profound humanitarian invention has evolved into an estimated $2 trillion industry, with a handful of platform companies vying for control. This has led Berners-Lee to confess that the web has lost its original egalitarian spirit. At the 2017 World Economic Forum, in a planned conversation with a handful of colleagues, ourselves included, he plainly told us that his creation has not yet become what it was intended to be. More recently, in an article for *Vanity Fair*, Berners-Lee was more to the point, admitting that the web has "failed instead of served humanity, as it was supposed to have done."[15]

While it is commendable that many tech titans have indicated they aim to ensure that 7.5 billion people can access the web and connect to one another by 2020, a conflict of global proportions has arisen: Many of these behemoth corporations are subject to shareholders and market cap obligations. In other words, they must monetize the web by converting it into a private network in which their users become their products. The shareholders and board members of these companies are far less interested in helping you make more friends than in how to best capitalize on your social data graph. The reason why many of their services are free or very cheap is because you pay them with the traits of your humanity, which are sold to advertisers for everyone's profit but yours. While we are finally wising up to this reality, there is much work still to be done to untangle the influential web we've wrapped ourselves in. The book in your hands is an important beginning.

Over the previous three industrial revolutions, humanity employed water and steam to mechanize production, then electric power to create mass production, and, finally, electronics and information technology to automate production. The Fourth Industrial Revolution has been growing from the third for the last half century, and it is characterized, according to renowned German economist Klaus Schwab, as "a fusion of technologies across the physical, digital and biological worlds."[16]

This fusion is a highly promising prospect if deployed properly. Global problems that have remained for decades, such as getting clean water and curing cancer, are now solvable in the present. Global mandates that have seemed unattainable for decades, such as universal education, are now a reality in our lifetime. But as it stands, the blurring of spheres that Schwab describes has largely led to the abuse of humanity's resources and the loss of our grip on the future.

The essence of the human spirit and the hope of humanity is freedom: the freedom to be ourselves, to express our personal convictions, and to become the best versions of ourselves that we can become. In truth, we are more than human beings; we are human *becomings*. What we collectively become writes the script for our world's future. Is technology enabling us to collectively become our best selves?

The precarious marriage of humanity and technology leaves us each with an existential question. It's staring us in the face and stirring inside of us. Do you want to merely survive, or do you want to truly thrive? Thousands of years of human history provide an obvious answer that sets us apart from every other life-form: Humans long to thrive, to progress, and to become better than we are now.

How we do this today is new territory. But we still possess within us the most innovative tools on the planet and, collectively, the most potent force for global progress. The opportunities are here, or very near, for the taking. Which ones should we take? Let's look into that question more specifically, on a topic-by-topic basis from here on out. Let's begin the discussions we must have. And then, let's begin forming our collective, proactive answers.

2

FRAMING OUR BEST FUTURE

If our intention is to ensure that local, national, and global technological advancements are each governed by a humanity-first objective, then how should we proceed? Do we simply allow inventors, innovators, and developers to create away? Do we allow elected leaders to decide what takes priority and what takes a back seat? Do we restrict our developments to those that can be proven to improve humanity? The answer to these last three questions is yes.

There must remain an unbridled freedom for those who create. This mind of ours is a marvel. And when collaboration is allowed to flourish, the results are more profound than what any algorithm can offer. If we do not allow creative freedom to reign, we risk missing the solutions we desperately want and need. But there must also be a sense of direction to the advancements we pursue—not only from an individual perspective but also from a global standpoint. The reality is that creative freedom always gives rise to risk—namely, the risk of following a perilous path that looks safe at the trailhead.

In the last chapter, we offered a startling example of creative freedom gone very wrong. This might seem like an outlying illustration, a once-in-a-millennium mistake, albeit a fatal and irreversible one. The truth is that the risks of creative freedom in technology have always been around. We've tiptoed on the fence of major catastrophe before, at times falling into failure and, at other times, dropping into significant success.

Early computer technology was quite inhuman and largely misunderstood by the masses. Its usefulness remained in the obscurity of the programmer's mind until Jobs and Wozniak created a human-centric approach called Apple. I think we can agree that this was an improvement in the marriage of humanity and technology. In a general sense, the introduction of Apple closed the divide between human and machine and elevated the use of computers to a practical, everyday level, which continues expanding today.

Yet, other attempts at elevating the human condition through technology produced mixed results, at best. We already mentioned the progression from X-rays to nuclear weapons using the same basic technology. But there was also the progression of automobile engines to automatic weapons, both using combustion technology. In the U.S. alone, we've seen the profoundly negative effect automatic weapons can have. The point here isn't to argue for or against gun control; the point is simply that the proper marriage of humanity and technology isn't crystal clear. Nonetheless, it is still important to try to get it right.

We can't convince every last earthbound innovator to think of the broadest implications of their creations (although we wish we could). And we can't convince every global citizen to think humanity-first with every action and investment of resources (again, this doesn't diminish our wishes). However, we can offer a broad and broadly accepted framework to serve as a guide for our primary objectives in this human-tech marriage—and to ensure that we are protecting, promoting, and advancing the goals of humanity, not suppressing them.

The remaining chapters of this book will use Maslow's hierarchy (see Introduction) as that framework. While we clearly acknowledge our debt to Maslow, one of the most celebrated psychologists in history, we do also believe that the application of his hierarchy to the human-tech marriage is a new perspective. Whether or not that's true, it can serve as an inspiring, simple guide for the sort of deliberate decisions we all must make today— on the topics that matter most to our future.

A word on Abraham Maslow. He believed that psychoanalysis focused too much on sick people and not enough on healthy people. He also believed that behavioral psychology, at the time, didn't allow enough distinction between human behavior and animal behavior. In response, he greatly contributed to humanistic psychology, which gained influence for its appreciation of the inimitable value of the human experience.

Maslow's main contribution to humanistic psychology (and psychology in general) was his theory of motivation, which focused on how humans meet their most important needs.[1]

Maslow described this hierarchy of needs as being made up of five main categories: physiological, safety, love, esteem, and self-actualization. These categories are arranged in a pyramidal manner, with physiological needs making up the bottom of the pyramid, and self-actualization, the top. He initially described the interplay of this arrangement in a fairly linear manner; humans, he said, had to meet their most basic needs (like food and water) before they could focus on the needs above them (like esteem or safety). He originally described this natural prioritization of needs as a "prepotency" that we are born with. However, Maslow later retracted that statement after further studies, concluding that humans could simultaneously pursue the meeting of needs up and down the hierarchy—while maintaining a basic survival instinct, which keeps us more focused on solving dire physiological needs than, for instance, our education or fitness.[2]

Maslow explained the hierarchy this way:

1. The **physiological** level includes needs that keep us alive, such as food, water, shelter, warmth, and sleep.
2. The **safety** level includes the need to feel secure, stable, and unafraid.
3. The level of **love and belongingness** addresses the need to belong socially by developing relationships with friends and family.
4. The **esteem** level includes the need to feel both (a) self-esteem based on one's achievements and abilities and (b) recognition and respect from others.
5. The **self-actualization** level addresses the need to pursue and fulfill all of one's unique potentials.[3]

Maslow's hierarchy of needs remains a major aspect of modern psychology and has been adapted for use in numerous other industries. The marriage of humanity and technology is perhaps the most important arena for this sort of adaptation.

If human motives are governed by the meeting of human needs, shouldn't we, as the body of humanity, use our partnership with technology to advance every human's ability to meet his or her needs—from our most critical requirements for survival to our desire to accomplish something that transcends our individual experience?

Maslow divided the five-stage model into what he called deficiency needs and growth needs. The first four levels are often referred to as deficiency needs (D-needs), and the top level contains growth or "being" needs (B-needs).

D-needs arise due to deprivation or lack, and we are highly motivated to meet these when they are unmet. The longer they go unmet, the more motivated we are. You've been famished or parched before—what happens? It becomes increasingly difficult to work, have a conversation, or do anything other than find something to eat or drink. This is the basic effect that Maslow's research identified (though the meeting of a lower need doesn't always precede or preclude the pursuit of a higher one).

B-needs don't stem from a lack of something; rather, they come from a desire to grow as a person. These needs are what set us apart from every other living thing in the universe. We don't just want to survive, we want to thrive—including living lives that positively impact someone other than ourselves.

Every person has the ability and the desire to move up the hierarchy toward the best version of themselves. Unfortunately, progress is often disrupted by a failure to meet lower-level needs—whether due to poor personal choices or to circumstances an individual can't control, like being orphaned or widowed, stuck in poverty, or born in a technologically underdeveloped nation. Helping those in the latter category overcome such circumstances ought to be the focus of humanity's surplus of resources, particularly those coming from technological advancements.

Ultimately, life causes most of us to fluctuate between levels of the hierarchy. However, the desire of every human on the planet is to reach a point when he or she can focus on meeting the needs of self-actualization, eventually using his or her daily experience to transcend personal needs and address the needs of others.[4,5]

In this light, our recommendation is straightforward: Let's use Maslow's hierarchy as a general guideline to ensure that our technological efforts meet important human needs—and that certain needs are not overmet at the expense of others, especially those needed for basic survival. The goal of humanity is neither to return to the Stone Age nor to fly in autonomous taxis over cities of starving people.

A better future world is within reach. The tools are available—within us and around us. Consensus is attainable where it matters most. But we must agree to wise guidelines, allow creative freedom to flourish within them, and then apply ourselves individually in supportive and innovative roles. Maslow's pyramid provides a simple, human-first framework for the technology we create, invest in, and adopt collectively. It shows us where we should focus our inventive and innovative activity, to ensure we are meeting important human needs, and it provides a sense of accountability, to govern a holistic global effort.

The remainder of this book will focus on some of the most important technological efforts throughout the world that fit within the guidelines of Maslow's pyramid. That is to say, they all meet human needs that allow us to thrive in greater ways in the following main categories of development: water, food, security, health, jobs, money, transportation, communication, community, education, government, and innovation.

If we were to place these critical areas of development into the construct of Maslow's hierarchy, it would look like this:

The transHuman Code

Our objective is a better world for all of humanity, and this is how we stay on the path. With our relationship to technology firmly established, the real question lies in how that relationship will be used and where it will lead us. Technology can be humanity's best resource, so long as we remember that *we* use *it*—and that we are the axis around which its use revolves.

In the end, our highest goals ought to be sufficiency for all and individual freedom to pursue abundance. If we don't begin to pursue the solutions humanity needs, there is a chance we will be too late in certain industries, even a year from now.

"The way to look at the world is 700 million people or so on the planet have a rich lifestyle," explains technologist and investor Vinod Khosla. "Seven billion people want it. Can we do ten times as much of everything the same way? The obvious answer is no. Technology . . . is the only thing that can multiply resources . . . It is probably the main way we will get to getting seven billion people the kind of lifestyle they'd all want."[6]

Let's begin this work together.

3

WATER

Meda, Ethiopia, is 150 miles southwest of the hottest place on earth, an abandoned outpost called Dallol, where temperatures reach 180 degrees Fahrenheit above ground, and hotter below—where pools of acid water boil inches beneath the surface. While Meda's weather is sweltering year-round, it's more forgiving than Dallol, and thus inhabitable. But the people living there must exist without a general store, power, or running water. In his eye-opening book Thirst, Charity: Water founder Scott Harrison tells the moving story of traveling to the village of Meda in 2014 to understand why a local 13-year-old girl named Letikiros took her own life.

Letikiros lived in a simple dwelling of mud and rock with her mother, sister, and new husband, a handsome priest's servant named Abebe to whom her mother had arranged her marriage. Life in Meda is a challenge, especially for the young girls like Letikiros, who, from the time they are about eight years old, are given the responsibility of collecting water every other day. The nearest source is Arliew Spring, a two-hour walk down a steep cliffside path strewn with scree, where villagers have fallen 700 feet to their deaths. At the spring, girls from Meda wait in line to collect water the color of chocolate milk, which trickles through large boulders covered in baboon excrement. Letikiros made this trip three times a week, returning home with five gallons of dirty water, the equivalent of three toilet flushes or a two-minute shower, in a developed country. When her days were not spent collecting water, she attacked her schoolwork vigorously—with her eye on one day using her education to help her village. By her 13th birthday, she'd completed the equivalent of third grade.

Harrison describes meeting Letikiros's best friend, Yeshareg, with whom she'd walked to Arliew Spring on the day she died.

"She was different," Yeshareg said of her friend. "Letikiros always dreamed of a better life for us. She would talk about leaving Meda one day, and helping to bring back health care, water, better education."[1]

Yeshareg recounted that the two had left before dawn that fateful morning, skipping breakfast to get to the spring early and avoid the long line. By noon, they'd filled their clay pots with five gallons of water each, strapped them to their backs with weathered rope, and begun the two-hour walk home. They parted ways at a fork in the path around 3 p.m.

"It was the last time Yeshareg saw Letikiros alive," writes Harrison. "Somewhere along the path, she stumbled. Maybe it was hunger, coupled with the forty pounds she had on her back [more than half her body weight]. Perhaps her legs failed her, or she tripped on a rock. All I knew for sure was that when she went down, her pot smashed against the ground, shattering into tiny pieces. The precious water she'd spent 10 hours collecting was gone in an instant, sucked up by the thirsty ground. Shortly afterward, a village elder was passing by when he saw Letikiros's limp body hanging from the branches, and the pot nearby, broken into pieces. He wailed in grief."[2]

After speaking with those who knew her best, Harrison learned that Letikiros had surely been overcome with the disappointment and shame of failing her family, who desperately needed the water. The 13-year-old slipped the empty rope around her neck and over the branch of a withered tree and hanged herself.

Letikiros isn't alone in her desperation—663 million people worldwide don't have the clean water they need to survive. To cope, many drink from contaminated puddles and streams—the very sources that make waterborne disease the world's leading killer and the cause of death for 4,000 children every single day.[3]

In developing countries, it's often the women and children who suffer most. According to the Centers for Disease Control and Prevention, African women spend 40 billion hours a year walking for water, and 20 percent of primary school–age girls are absent from school, often because there are no sanitation facilities for girls approaching adolescence, or because the responsibility for collecting water often falls to the young girls in a household, making it impossible for them to attend school on a regular schedule.[4] All the while, clean water is often flowing—like liquid gold—in untapped rivers just below their feet.

The Quest to Quench Our Thirst

Should it be, at a time when we possess the technology to launch rockets into space and dive submarines to the bottom of the deepest oceans, that we cannot find a way to ensure every human on the planet has clean water on a daily basis?

Today we must ask: What are the barriers keeping families like Letikiros's from simpler access to clean water? And what advances in the technology sector can help deliver clean water to the world's nearly 700 million people who don't have it?

There are three primary objectives that must be advanced and scaled in the quest to provide clean water to everyone on the planet:

1. Locating and accessing water
2. Treating and cleaning water
3. Reusing water

Locating and accessing water has often been the bottleneck in the water supply chain. This is changing. Findings by Stanford researchers Rosemary Knight, Jessica Reeves, Howard Zebker, and Peter Kitanidis could eliminate the costly guessing game of drilling wells. Until recently, the only way to assess the state of water tables was by installing monitoring wells and then comparing that information with existing well water levels. But an in-depth study of water levels in the American West found that the data was not 100 percent reliable, either because available data is outdated and of varying quality or because not all well data is shared by the owners. This is the case in a *developed* nation. Imagine the limited and unreliable data available in *developing* nations, where funds may not be available to scatter and keep tabs on monitoring wells in arid, less-populated regions.

The group of Stanford researchers joined forces with Willem Schreüder of Principia Mathematica, Inc., and turned their attention from monitoring wells to satellites. These satellites emit electromagnetic waves, enabling them to gather measurements and keep tabs on tiny changes in surface elevation. The technology, known as Interferometric Synthetic Aperture Radar (InSAR), has historically been used to chart data on potential earthquakes, landslides, and volcanoes. The researchers hypothesized that InSAR could *also* be used to establish clearer clues about groundwater and, thus, provide more reliable information about when and where to dig, reducing the cost of drilling multiple wells.

According to a news report by Rob Jordan on Stanford's website, "With funding from NASA and the School of Earth Sciences, the researchers used InSAR to make measurements at 15 locations in Colorado's San Luis Valley." When they compiled their findings, they discovered their data matched the data that had already been compiled from nearby monitoring wells. In other words, their hunch was correct. Another research team, using similar methods, successfully mapped the groundwater levels of more than 500 rivers, lakes, and flood zones in the Amazon basin.[5] "If we can get this working in between wells," Reeves explained, "we can measure groundwater levels across vast areas without using lots of on-the-ground monitors."[6]

"Just as computers and smartphones inevitably get faster," writes Jordan, "satellite data will only improve. That means more and better data for monitoring and managing groundwater. Eventually, InSAR data could play a vital role in measuring seasonal changes in groundwater supply and help determine levels for sustainable water use."[7]

Discovering and tracking groundwater via satellite could be a major breakthrough in the discovery and consistent provision of water in arid environments, perhaps even bringing the affordable and accurate drilling of wells to places like remote villages in Ethiopia. But it's not enough for a handful of scientists and researchers to advance this important application of technology. We must join forces to improve the technology and scale it around the globe—in particular, to the areas where the 700 million people without clean water live. And then we must continue innovating further, because one tool or application is rarely enough to solve a critical global problem.

Case in point: Letikiros's village of Meda is not a candidate for the new application of InSAR. In *Thirst,* Harrison explains that field studies determined the only reachable groundwater in the area would require "a massive pipe system costing upward of half-a-million dollars . . . It's one of the hard realities we deal with all the time."[8]

The good news is that water isn't just flowing in rivers under our feet. In some environments, during particular seasons, evaporated water is floating in the air all around us. New technologies are dedicated to identifying that water, capturing it, and delivering it to communities where water has, up until now, remained a scarce resource.

One such tool is called the WaterSeer, a device that uses the surrounding environment to pull water from the atmosphere and guide it into an underground chamber, where it is cooled and stored. In good conditions, the WaterSeer can gather 10 gallons of water from the air every day.[9]

In other parts of the world, huge net formations catch the moisture from fog. This moisture drips into collection trays, making the accumulation of water immediately accessible. This method was first developed in South America, but the largest such project is on the slopes of Mount Boutmezguida in Morocco, where 6,300 liters of water are collected every day.[10]

Where these technologies have been successful in increasing the water supply, other problems can arise; addressing these problems is just as important as locating the water in the first place. For instance, who will control the water supply in any given area? The government? The private owner of the land beneath which it flows? A public corporation? And who will maintain it? In locations where water was at one time scarce, how can farmers learn to irrigate their crops in responsible ways and at proper levels, so as to not prematurely run the supply dry?

One possible answer to the last question is to incentivize farmers who use solar pumps to sell excess power back to the grid, increasing their income, adding to the government's energy reserves, and, ultimately, conserving water by curbing their use. There's an ancillary benefit to going solar—one that is perhaps equally significant. The International Water Management Institute estimates that introducing solar power to India's 20 million irrigation wells could bring down carbon emissions by 4 to 5 percent per year.[11]

Introducing technology to find and distribute clean water is not the only way to help the nearly 700 million people who do not have it. In some instances, dirty or contaminated water, if treated properly, can transform communities and eradicate disease. According to the World Health Organization, "Diarrhoeal disease is the second leading cause of death in children under five years old," taking the lives of more than half a million of them each year. It is contracted from a lack of clean drinking water and basic sanitation, making it both preventable and treatable.[12] Thus, treating existing water supplies is one of the leading strategies for water provision. Historically, though, the cost of treating contaminated water, or desalinizing ocean water, has been prohibitive.

Recent advances in technology are mitigating these costs. For example, one of the unforeseen side effects of the highly controversial hydraulic fracturing industry has been the demand it has created for mobile water treatment facilities. Massive corporations are making large investments in the creation of portable reverse osmosis units, which will enable companies to treat high volumes of water, extracting gas and debris. As companies pay for technology that treats water in increasing volumes, it will be possible to move away from the current system of massive, centralized, and expensive treatment centers.

According to *The Guardian*, researchers in India devised another solution to cleaning contaminated water: nanotechnology. Microbes are removed from the water using composite nanoparticles that destroy contaminants. The cost is only $2.50 per family per year. These advances illustrate that low-cost water purification is on the innovator's horizon and could finally be commercially viable.[13] However, we must also question whether mobile water purification needs to be commercially viable or necessary—in a booming industry like hydrofracking, for instance—in order for it to be resourced or investment-worthy.

We can find potable water. We can see the beginning of a future where treating dirty water is affordable and portable. But how can we better recycle the water we've already used? Emory University's Water-Hub employs an adaptive ecological technology that recycles "sewage to usable water for heating and cooling campus buildings," which reduces the Atlanta-based university's water footprint by 40 percent.[14]

Perhaps a new frontier is emerging in the recycling of wastewater, making it possible for areas of the world desperate for clean water to sustain themselves—once water has been found and purified. To realize that possibility, the best solutions must be used together, and they must reach beyond the campuses of affluent American universities. These are challenges not only of innovation but also of value and prioritization.

Water Technology Introduces New Challenges

These challenges in supplying the world with clean water are being overcome, and technology is paving the way to new answers. Potable water is being found more quickly and efficiently. The treatment of contaminated water is becoming more affordable. And new approaches are allowing for the reuse of wastewater. But the introduction of technology into procuring, treating, and reusing water hasn't only brought answers; it has also raised new and increasingly important questions.

For example, in 2006, hackers gained access to computer systems at a water treatment plant in Harrisburg, Pennsylvania. They used an employee's laptop, compromised by way of the Internet, to install a virus and spyware in the plant's computer system. The attackers operated from outside of the U.S. and were not specifically targeting the water plant (they were simply using the laptop to distribute emails and other electronic information), but had they been interested in controlling or manipulating the plant's systems, they may have been able to cause serious damage, perhaps by raising the chlorine level and making the water dangerous to drink.[15] And as recently as 2018, the U.S. blamed the Russian government for an ongoing campaign of cyberattacks aimed at U.S. infrastructure, including water management sectors.

Developing nations, barely able to afford these water treatment technologies, simply do not have the funding to properly secure them, providing soft targets for hackers.

Michael Deane, executive director of the National Association of Water Companies, explains how the evolution of computer-based management systems has, on the one hand, improved the reliability and quality of water services but, on the other, has increased the possibility of targeted or accidental cyberevents that could disrupt the water supply. He concludes:

> In the drinking water and wastewater sectors, a cyberattack could hone in on four different threat vectors: chemical contamination, biological contamination, physical disruption, and interference with the highly-specialized computer systems controlling essential infrastructure, known as Supervisory Control and Data Acquisition (SCADA) systems.[16]

New technologies always invite new and more pressing questions. This is certainly true with water-related technology. Are we willing to address these new questions before the answers are forced on us? Our collective answer to that question alone could save 700 million lives.

The utilization of a new technology may at first seem overwhelmingly positive; where new ways of delivering drinking water to underserved communities is concerned, it's vital. But utilizing these new technologies is often like toppling the first domino. We must, therefore, look all the way to the last domino, prepared to examine and answer all of the questions and challenges that could arise in between the two.

When the country of Spain began a campaign to desalinate its water, it wasn't seen as wholly necessary. Today, all of their drinking water is provided by this source—and it's a good thing, because Spain is as dry as the Sahara now. The decision to begin desalinating turned out to be a critical one that was made early enough.

In the same spirit, but on the reduction side of the equation, China is undergoing a toilet tech revolution (new toilets that you don't flush), which translates to a much more efficient use of the country's water. In Switzerland, every supermarket product has a label that indicates how much water was needed to manufacture the product. The country is raising consumer awareness about the importance of using its water wisely.

In contrast, in the fashion industry, where water technology has yet to be widely embraced, a cotton shirt requires 2,700 liters of water to make.[17] Multiply that by the number of cotton shirts in the world's closets and you can see that a more thoughtful application of technology can make a big difference.

While we are moving in the right direction where clean water is concerned, it is *slow* movement. Unless we're talking about the latest

LifeStraw-type product (which enables hikers to drink directly from open water sources), supplying water isn't a particularly sexy topic. The latest InSAR study won't be featured in *Vogue* or *The Wall Street Journal*. The tragedy of Letikiros is unlikely to find its way into a Hollywood screenplay. Unfortunately, we live in a world where top-of-mind often commands top dollar. Water provision is not top-of-mind at the moment—but it should be, if only because we have the resources to solve this problem now.

So where and how should we begin?

First, we need to enter into a multinational, multidisciplinary discussion that seeks to wisely address the questions mentioned in this chapter:

1. What are the barriers keeping families like Letikiros's from simpler access to clean water?
2. What advances in technology can help deliver clean water to the nearly 700 million people around the world who don't have it?
3. What *more* can we do to improve and scale . . .
   a. Locating and accessing water
   b. Treating and cleaning water
   c. Reusing water
4. How can we ensure that the water supply chains we create for underserved areas are not abused?
5. Which people and/or entities should fund, control, and manage the new water supply chains that are created?
6. How can we redirect investments from noncritical ventures to critical ones—like the supplying of clean water to all humanity?

Second, from our discussions on the topics above, we each must, as individuals, determine immediate courses of action to be taken where the investment of our time, money, and ingenuity are concerned—not over the next 5 or 10 years, but now.

1. If you write code, what HI must be written into the AI technologies of water supply so that lives are protected and investments in the supply chain are not wasted? Can you or someone you know do it?

2. If you innovate, what other water supply solutions should we consider? Who do you know that can provide them?

3. If you build or manufacture, what materials might make components of the water supply chain more viable?

4. If you finance, how can we best fund our efforts to supply every person on the planet with clean water? Should the answers come only from the nonprofit sector?

Each of these questions must be discussed openly and purposefully if we are to use technology to end the thirst that plagues 10 percent of humanity.

4

FOOD

We live in an age where anything is possible, even the engineering and production of food in laboratories. Bioengineered plants, replicated beef and fish proteins—nearly every food can be created in a laboratory these days. But does that mean these sorts of possibilities should drive every solution to global food security? From a cost perspective, could they? Hardly. Sometimes human ingenuity is the best technological application.

We sat down with Dr. Evan Fraser, the director of the Arrell Food Institute at the University of Guelph, and he shared the story of some Nepalese farmers in a small agricultural community who were struggling to provide food for their families. Despite the number of hours spent in the fields and the large expenditures of human capital, crop outputs were low. To achieve anything resembling food security, they needed help boosting yields.

Enter Dr. Manish Raizada.

Dr. Raizada, a Stanford-trained molecular geneticist specializing in plants, learned of the Nepalese farming problem and figured he could help. He made his way to the community, wondering if the application of genomic technologies might boost corn production. But once he and his team put their boots on the ground, once they interviewed the farmers and compiled their notes on the local planting methods, they discovered the predominant problem with the Nepali crops. It wasn't that the farmers didn't have a high-yield seed. It wasn't even that they lacked more technical means of irrigation or fertilization. They didn't need more advanced combines. The predominant issue with their crop production was a human problem.

Dr. Raizada's team discovered that the vast majority of the Nepalese farmers in the particular region weren't familiar with best planting practices. Instead of preparing the soil and planting in evenly spaced rows, they took a more traditional approach, scattering handfuls of seed in the dirt. Knowing that this planting method led to plant crowding, which didn't allow the plants access to proper nutrients, moisture, and sunlight, Dr. Raizada's team set up some test plots. They prepared the ground and planted their seeds—the same seed used by the locals—in evenly spaced rows. There was no reported change in irrigation or fertilization methods. No reported change in the amount of energy used to work the crops. And when the harvest came, Dr. Raizada's test plots produced between 25 and 40 percent more corn. The yields had increased, all because simple agricultural best practices had been used.

Raizada's team could have employed the latest technology to increase yields. He could have procured the latest genetically modified seeds, which would have produced an abundant harvest, regardless of row spacing—but he didn't. Instead, he and his team resorted to simpler methods. And having proven their hypothesis—that proper spacing would increase production—the team created a simple tool kit that made it easier for the local farmers to plant in rows. In that kit were two sticks with an attached string, to mark lines on the ground, and a hollowed-out stick for dropping one seed at a time, which meant the planters—many of whom were women—didn't have to crawl on their hands and knees tediously. It was simple and affordable, a distinctly human application of technology. It was a methodology that could be used by even the poorest of the Nepalese farmers.[1]

The Fragile State of Food Security

"The world is at a critical inflection point with respect to food," said Dr. Fraser.[2] And that's not hyperbole. As of 2017, the world population was estimated at 7.5 billion people, and projections indicate that the earth will be home to some 9 billion people by the year 2050. In order to feed that entire future population, we will need to increase our current food output by roughly 70 percent, and output from the developing countries will need to double.[3] Fraser's front-line research indicates that in order to keep up with food demand, the world's agricultural systems will need to produce more food in the next 50 years than they have over the last 10,000 years.

But before we address the problems of tomorrow, we have to take a hard look at the problems of today. In countries like the United States, the United Kingdom, France, and Russia, the per capita daily caloric supply is over 3,400 calories. However, in many places across Africa and South Asia, the daily caloric supply is under 2,500 calories per person.[4] As of 2012, the Food and Agriculture Organization of the United Nations reported that 870 million global citizens were chronically malnourished.[5] And while so many are malnourished, it's reported that approximately 2 billion people are obese.[6] Despite extremes in malnourishment and obesity, over one-third of the world's food is still wasted.[7]

The headline here is that it's not how much food we produce that will determine whether or not the future is well-fed. It's how we produce it, what we produce, and the extent to which it is equitably distributed. Those are the components that we have to be deploying technology for. Not just boosting production.

Fraser shares an illustration of this, to illustrate it in a slightly unfair but accurate way. "A couple of years ago, I was leading a group of students on a tour across the Midwest [of the U.S.] and we stopped at [a major] seed company for a tour of the headquarters of their genetics work. We talked to the head geneticist responsible for corn reaping and corn modification. And he said proudly, 'Our first three priorities over the next few years are to boost yields, number one, to boost yields, number two, and our third priority is to boost yields. And we're going to move from 300 bushels per acre up to 600 bushels an acre.'

"Now most of the corn in the Midwest is produced either for livestock food, to drive the cost of fast food down, or for simple sugars, cornstarch, and corn syrup. The guy that was talking to us weighed 350 pounds, and was waving a cola in one hand and a hot dog in the other. So, the last thing this world needs is more corn in the Midwest."[8]

The numbers are daunting, the challenges, many. How can we increase food production—especially in regions where access to irrigation, fertilization, capitalization, and crop production technologies are limited? And what technological advances might help reduce the pressure on the food supply in more developed countries?

There are technological methods for increasing food production, of course, and in our age of advancement, those methods are increasingly being applied. Just as in every other industry, the digital revolution has come to agriculture. The same technologies producing the Internet of

Things (IoT) are transforming our food and farming systems. This revolution has been disruptive from the farm to the table—and in many regards, these disruptions have been positive. For instance, drones (paired with satellite technology) are now finding the best locations to produce crops, along with the best soil conditions and access to water, helping farmers increase production while substantially reducing water consumption. Additionally, sophisticated soil and thermographic sensors, together with cutting-edge software platforms, provide growers vital information about soil and plant moisture levels, enabling them to make informed and precise decisions about irrigation frequency. These methodologies lead to better plant health and overall yields, and ultimately provide the consumer with more nutritionally dense foods.

But though connected farm equipment might lead to increased food yields—an inherently positive outcome—IoT advancements alone won't alleviate global hunger. Why? There are billions of people without access to the technology.

There are certainly technological solutions to the problem, solutions that do not rely on IoT devices or more advanced machinery. For instance, though highly controversial, genetic modifications can be used to help solve food disparity issues. By removing the genetic material from one organism and inserting it into the permanent genetic code of another, the biotech industry has created an astounding number of organisms never before seen in nature or on a plate, including both potatoes and corn with bacterial genes, tomatoes with flounder genes, fish with cattle-growth genes, and "super" pigs, with human-growth genes. There are thousands of other altered and engineered plants, animals, and insects now being patented and released into our environment and food supply—at quite an alarming rate.[9] But modification isn't all bad; in fact, it can be very beneficial.

We've created supercharged foods that grow under less than optimal conditions in remote areas, where farmers have little (if any) access to modern technology. For example, in famine-prone regions of Africa, golden rice—a genetically engineered, nutrient-rich grain—has been grown to help alleviate famine and vitamin A deficiencies. And golden rice is just one of millions of genetically modified crops that have the potential to grow in less than ideal circumstances, resist drought and insect infestation, and provide the community with increased nutrients.[10]

Sounds great, right?

Although modern genetic modification could be used to create foods that would end global hunger, these technologies are often applied to less nutritious crops, like corn. In fact, some suggest that less than 5 percent of the world's non-corn crops benefit from genetic modification. Why has this technology been primarily applied to corn? Consumer demand.

Though corn is an imperative dietary staple for those around the world, and though it wouldn't be difficult to imagine a genetically modified and nutrient-dense corn, the majority of the world's corn isn't produced for direct human consumption. Why? Demand for meat and sugar are at an all-time high. A 2013 article in *Time* magazine indicated that "about 30 percent of the world's total ice-free surface . . . is used not to raise grains, fruits, and vegetables that are directly fed to human beings, but to support the chickens, pigs, and cattle that we eventually eat."[11] And the majority of the rest is used to create sugar-based products. Though these sources of food are calorically rich, many global health challenges result from the overproduction of sugar and meat and the underproduction of nutrient-rich vegetables and grains.

Modern agriculture produces two times the recommended servings of carbohydrates (six servings instead of three) and less than half the servings of fruits and vegetables needed (less than five servings instead of ten). We're overproducing fatty meats and sugars, mostly for consumption by the wealthiest countries. The result? Chronic malnutrition, protein deficiency, and an increase in diseases like diabetes, obesity, and hypertension. And this is to say nothing of the deterioration of our agricultural microbiomes.

The application of genetic modification to satisfy our demand for the wrong kinds of foods is responsible for short-term caloric gains, but in the long run, it's harmful to human health. What's more, it centers those short-term gains in developed societies, which can afford the application of those technologies, ultimately leading to disparity in global caloric distribution. For instance, although it is estimated that the global food supply would allow each person 2,850 calories per day, hunger still plagues parts of the world.[12]

The Risks of a Technological Agricultural System

The use of technology to increase food production can also exacerbate the already heavy impact of agriculture on the environment. According to current estimates, livestock production uses one-third of the world's fresh water, and some indicate that each cow grown in the United States requires 20 pounds of feed for every 1 pound of edible meat produced.[13,14] (Consider if these agricultural inputs could be shared with the world.) But the environmental impact of cattle production in underdeveloped countries is high too. Some estimate that the developing world produces 75 percent of the world's cattle-related greenhouse gas emissions.[15] And according to a 2006 report by the Food and Agriculture Organization of the United Nations . . .

Overall, livestock activities contribute an estimated 18 percent to total anthropogenic greenhouse gas emissions from the five major sectors for greenhouse gas reporting: energy; industry; waste; land use, land use change, and forestry (LULUCF); and agriculture.[16]

And this is to say nothing of the greenhouse gases created by the use of farm equipment or the application of fertilizers.

As we apply more and more technology to meet rising food demands, our agricultural carbon footprint will continue to expand. But technology can help us address some of those problems, too. In recent years, connectivity and modern robotics have helped reduce our footprint. Modern tractors with GPS-enabled robotic sensors can locate their position in the field and ensure that the right seeds are planted at the right interval. What's more, these machines give the precise measurement of fertilizer needed by the plant, eliminating waste and reducing carbon emissions from overfertilizing.

Aside from these advancements, robots are increasingly employed in nursery planting, thinning and pruning, picking and harvesting, and even milking cows on dairy farms. So, in addition to saving natural resources, these advances theoretically conserve human resources. In the best instances, this should allow farmers to concentrate on improving overall production yields. That stated, this same use of robotic technology may eliminate jobs, thus creating greater income disparity.

In addition to environmental and labor risks, genetic modification and high-tech machinery may open our global food supply up to other dangers. For instance, the connectivity of farm equipment makes the threat of cyberattack by hackers a very real possibility. Smart tractors can be compromised; soil sensors, corrupted. Water distribution systems may be infected by malware. In short, farmers, co-ops, and agricultural transporters who don't have cybersecurity at the forefront of their minds are at risk—and even then, they'll be no match for sophisticated hackers. The many technological entry points of attack in modern agriculture, therefore, might jeopardize food production and distribution networks, adversely impacting food security.

And if all these risks weren't enough, consider, also, the issues surrounding personal privacy in the use of this kind of connected agricultural equipment. It's been noted, for example, that John Deere is taking farmers' data and creating agricultural decision-support tools and tractor algorithms based on it. And once those tools and algorithms are created, they'll likely be sold back to the farmer and the farmer's competitors, all based on data John Deere didn't generate in the first place.

Will Modern Technology Become the Answer to Food Security?

Technology may help improve production and reduce our carbon footprint, but so far, it doesn't seem to be alleviating the threats to food security. It doesn't seem to be creating a safer, cleaner, more sustainable source of food for all humankind. There's been no indication that we're on our way to increasing yields by 70 percent globally, or doubling yields in underdeveloped countries. So, though we may theoretically have the resources and technology needed to eradicate hunger and ensure long-term food security for all, the question remains: Will the theory ever become a reality?

Instead of simply using technology to increase production, what if we were to look at all the unintentional by-products of that increase—inequitable distribution, lack of nutrition, greenhouse emissions, elimination of jobs, cybersecurity issues—and apply technology in ways that reduced those potential problems? Some are doing just that, in both high-tech and low-tech ways. Consider, for example, a group of innovators in the meat industry who have their sights set on reducing the environmental impact of meat production. How? By using technology to create a new sort of food product: meatless meat.

In August of 2017, Memphis Meats, a Silicon Valley start-up, secured $17 million in series A funding for its meat products, which are created through cellular agriculture, a process in which scientists grow meat from animal cells. And Memphis Meats isn't alone in its attempt to grow meatless meat. Its rival, Impossible Foods, creates its products from isolated plant proteins, amino acids, and vitamins, and the enthusiasm for its process of meat creation is off the charts. In August of 2017, Impossible Foods raised $75 million in investment capital, with that capital coming from the likes of Microsoft billionaire Bill Gates.[17] With capital injections like this, low-impact, sustainable meat production will be economically feasible in the not-so-distant future.

Still, although high-tech applications of technology can alleviate some of the issues with modern agriculture, their impact is relatively small. Add to that the fact that we're years (if not decades) away from the application of those technologies in the most impoverished places in the world—the places where food security is lowest. (It's true: Modern technological approaches are applied in ways that benefit populations with the most capital instead of ways that best serve the interest of all humans.) Given all of this, it's clear that the most basic human element—the element of human *need*—is missing from many of our discussions about agriculture.

In our interview with Dr. Fraser, he made this point very clearly.

"The picture of technology use I'm trying to paint here is not, 'Let's GM [genetically modify] some more corn. Do some genomic jiu-jitsu on corn in the Midwest. Boost the yields and then end up with more corn syrup on the planet.' I'm saying let's be far more nuanced, or critical, or empathetic to how we're going to use technology in order to produce nutrition, safety, equity. There's always a way for a smart scientist and a smart farmer to work together to come up with a cool solution."

Perhaps this is what made Dr. Raizada's approach to agriculture and technology so important. He considered the need, and though he could have deployed technology to increase crop yields—whether through modern genetics, robotics, software, or otherwise—he took a more human approach. He put boots on the ground, spoke with the local farmers, and examined their planting methods. He built relationships and looked for innovations to help increase production in relatively low-tech ways. He also assembled a team, and together, they applied the kind of human technology you can't program into a machine—empathy, connection, and human understanding—to the low-corn-yield problem of the Nepalese farmers. As they did, the solution presented itself.[18]

In this world of advancing technology, we may be tempted to believe that modern methods can solve any problem. But the disparity in the application of those methods—and the allocation of resources to the things people demand (versus the things they need)—render this belief naïve at best. So, in order to make headway in the area of food security, the smartest scientists need to work with the best farmers in the region to find culturally approachable, technologically appropriate, and cost-effective solutions to the problems that threaten long-term food security.

Dr. Raizada examined the landscape, and the problems were evident. In the same way, when we take a hard human look at the current status of global agriculture, the problems are evident: food scarcity and disparity, the loss of biodiversity, the creation of greenhouse gases, and cybersecurity threats to the food supply. There's no denying that both current and emerging technologies can solve some of these issues—yet, they simultaneously create another set of problems, with increasingly negative implications.

How do we create greater food security *and* reduce those potential problems? As we did with our examination of water security, let's enter into a more holistic conversation to answer these questions:

1. In what ways can we apply simpler human technologies to increase food security?
2. What advances in technology might be applied in developing countries to help provide higher-yielding, more nutritional crops?
3. What changes might developed countries be willing to make, and what resources might they be willing to shift to ensure greater food security and caloric distribution for the global population?
4. How can we reduce (or even eliminate) cybersecurity threats to our agricultural systems?

These questions may require sacrifices by the wealthiest among us. It may require a reallocation of resources away from crops *we* demand—crops such as corn and red meat—and toward systems that shore up our global food supply. It may invite further questions, too, such as—

- What am I doing to change the way I eat, to shift demand away from the wrong kinds of foods?
- How do my decisions affect my neighbors across the globe?
- What kinds of technologies might be used globally to increase food security, and how can I invest in or contribute to them?

When it comes to applying new technology to our agricultural systems, things have never been more exciting. And considering the world's rising population and global demand for better nutrition, the stakes have never been higher. Along with that, the ironies, tensions, and risks have never been more obvious or prevalent.

Whether or not the future human population is well-fed isn't really a matter of the quantity of food we're able to produce—it's a matter of the quality of the food and how equitably we distribute it.[19] In other words, as we deploy new technologies in the agricultural sector, we shouldn't begin the conversation by asking how we can increase yields but, rather, how we can address global human needs.

5

SECURITY

Two months before Drew received the first phone call from the police, his cyber double was already living the life of a millionaire. In a recent article for Bloomberg, Drew Armstrong explains that, using a driver's license with his name and photo, a complete stranger opened accounts at four separate banks in two days and then used those accounts to obtain a credit card from Bank of America, also under Drew's name. From there, the man spent a few nights living the high life at the exclusive Delano Hotel in Miami Beach; shopping at Whole Foods; and even selling an RV online, failing to deliver the imaginary vehicle, and then taking the $39,000 the would-be buyers wired him and sending it overseas.

There is video footage of this man in a Wells Fargo branch, posing as Drew and opening more accounts. The stranger, by obtaining a few important documents and numbers, had stolen Drew's identity, and for the next three years, Drew was left picking up the pieces. The man's financial crimes followed Drew from country to country, because they now shared a name—but not just a name, an identity. In the eyes of the financial world, they were the same person. Every time Drew left or entered the United States, he was pulled aside, his bags were searched, and he was often interrogated for up to an hour. Eventually, Drew was able to get a letter with a redress number from the TSA, allowing him to travel relatively unscathed, but in reality, that was the least of his problems.

It was an ongoing nightmare. For three years, Drew lived with a financially irresponsible version of himself out there, continually causing more chaos, while he filled out forms, talked to multiple departments at every bank, and searched for the right agencies and organizations to which he could issue his complaints. There was always more paperwork to be filled out or submitted. For example, in order to close the Bank of America credit card opened in his name, Drew had to send the bank the following:

- A signed statement saying he was who he said he was
- A list of the fraudulent accounts
- An affidavit filed with the FTC, swearing his identity had been stolen
- Copies of his driver's license, passport, and social security card
- A lease agreement and two phone bills
- A letter from the Justice Department and the criminal complaint he had filed with the police
- Bank of America monthly statements

It took all this to close one credit card account.

Eva Velasquez, CEO of the Identity Theft Resource Center, said that part of the problem is our desire for convenience: "We demand convenience over security. When you have an experience [like Drew had], convenience becomes less of a priority."[1]

Cybersecurity breaches are not only the concern of individuals—some of the largest corporations on the planet have seen stock prices plummet and executives resign after failing to maintain their data securely. Home Depot lost 26 million credit card numbers in 2014, a breach that eventually cost them nearly $180 million in compensation to those affected.[2] Three billion Yahoo user accounts were hacked in 2013, leading to 41 class action lawsuits.[3] And 117 million LinkedIn user accounts were compromised in 2012; the hacker was later found selling the emails and passwords for Bitcoin on a dark web marketplace.[4]

And then there was the Equifax breach: 143 million records containing names, social security numbers, birth dates, addresses, and driver's license numbers. CEO Richard Smith resigned immediately after news of the hack broke, stock prices fell by 35 percent (and remained flat), and the company incurred $87 million in costs having to do with the security breach.[5]

Online security is no longer solely the concern of cybersecurity experts. It affects individuals and organizations around the world.

The Loss of Privacy

It seems as if we hear news of this type on a weekly basis, as companies struggle to keep their security measures ahead of today's hackers and cybercriminals. The loss of privacy through identity theft is not uncommon: The U.S. Bureau of Justice Statistics estimates that 17.6 million Americans were the victims of at least one incident of identity theft in 2014.[6]

Yet, the threat of cybersecurity failures does nothing to dull our reliance on the Internet. In fact, as time passes, the Internet only becomes more and more embedded in our everyday lives. Everything in our new smart cities will have sensors, trackers, and microchips emitting information in real time, providing data about the status of anything happening in the city at that particular moment: the movement of a car or a person, the status of any of a thousand merchandising supply chains, or the whereabouts of a pet or a child. Anything circulating in the city will be immediately identifiable via a sophisticated network of cameras and sensors, sending real-time data to artificial intelligence clouds that will analyze it and compare it to behaviors that trigger complex alert systems.

Not only whole objects will be connected to the Internet: Their individual parts will be too, providing real-time performance and maintenance data. Connected cars will be analyzed through the cloud, even as they are traveling, to ensure that there is no security risk for passengers. Drivers will no longer be required; automobiles will have total autonomy.

This invasive tracking will constantly threaten to compromise the privacy of users. Still, the trend of tagging everything, connecting everything, and tracking it all shows no sign of abating, especially in places where security needs are greatest—places like public transportation hubs, government facilities, and areas of high population density.

With the increasingly ubiquitous nature of the Internet of Things— the interconnectedness of all we say, do, and think—how will we ensure that our privacy, our lives, and our possessions remain safe? Some users will start to define and create privacy cocoons, places where they can disconnect from the global surveillance system, places where they can vanish into some kind of obscurity—perhaps inside their cars or homes. Some employers will seek to create this kind of untracked space inside their offices, enabling employees to maintain a level of privacy that would otherwise be impossible to experience.

But these privacy cocoons will be the exception, not the rule. As more and more of us perform virtual check-ins everywhere we go, pay for goods using accounts set up on our cell phones, and drive cars with computer chips that are constantly being monitored, how can we retain any kind of privacy or online security?

The Internet Was Not Created with Security in Mind

Perhaps the heart of the problem is the fact that the World Wide Web was not originally designed to have any security: It simply was not a concern. The entire system was brought together as a method for connecting people on servers and had no reason for shielding them from each other, so privacy controls and security settings were not embedded in the original design.

Obviously, concerns are changing. Over the last 10 years, cybersecurity has moved from being a fringe topic, mostly talked about by online businesses, enterprises, and governments, to being the number one issue for CEOs, boards of directors, and *individuals*. One well-timed hack could cause a corporation to lose its entire valuation. One lost U.S. social security number could lead to the upheaval of an entire life.

To illustrate the immense nature of the problem, it's a fact that basically everyone will be—or already has been—hacked. How is this possible? Enterprises are still not protecting individuals against basic threats with simple measures like . . .

- Secure email
- Secure websites
- Encrypting communications
- Digital identities for employees
- Blockchain integration to decentralize data
- AI to prevent and analyze threats

So what lies at the heart of cybersecurity, and what are the ethics surrounding the continued development of technology?

At the Heart of Cybersecurity: The Human

TechRepublic's Dan Patterson, in a recent interview with Cisco's trust strategy officer, Anthony Grieco, made this interesting statement: "Humans remain the intractable cybersecurity problem. They also represent a cybersecurity potential solution . . . [H]umans are the challenge for cybersecurity and trust is one way to solve that problem."7

What he says is true: There are no cybersecurity breaches (as of yet) without a human instigator, and monitoring human behavior (i.e., keeping the wrong people out of secure online areas) is the primary concern of cybersecurity.

But how are humans also the answer to the cybersecurity problem? Perhaps developing increased levels of trust will create the kinds of relationships and environments that make it more difficult for cybercriminals to operate. For example, companies doing business in the 21st century must address customer concerns regarding online security, and as they do so, differentiating themselves from other businesses, they will begin to build trust with their customers and partners.

Later in the interview, Grieco shares that trust is built in multiple ways, such as having a vulnerability disclosure policy or being open about your security development life cycle. More advanced customers may ask more advanced questions, so implementing nondisclosure agreements may allow companies to be even more transparent with specific customers.

"In some limited instances, it may even make sense to go even deeper," Grieco says, "into a deeper relationship, a deeper partnership . . . [with] conversations about design and architecture and many of those sorts of things . . . The trend in this conversation is one that is more towards public disclosure. More towards openness and more towards transparency in all aspects of these businesses, because there's such a hunger from the marketplace to really understand what's going on in this space."[8]

Grieco goes on to say that the issues around cybersecurity are no longer an awareness problem. Everyone knows about the latest security breaches. What people are really concerned about is trust—the ability for customers and businesses, and everyone involved in between, to develop trust around their cybersecurity measures and to maintain it.[9]

And the question at the forefront of this conversation on trust is "How should we be navigating advances in artificial intelligence?"

The Ethics of Artificial Intelligence

In the summer of 2017, Eric Horvitz turned on the autopilot function of his Tesla sedan. Not having to worry about steering the car along a curved road in Redmond, Washington, allowed Horvitz to better focus on the call he was taking with a nonprofit he had cofounded. The topic of the call? The ethics and governance of AI.

That's when Tesla's AI let him down.

Both driver's side tires bumped up against a raised yellow curb in the middle of the road, shredding the tires instantly and forcing Horvitz to quickly reclaim control of the vehicle. Because the car hadn't centered itself properly, Horvitz found himself standing on the sidewalk, watching a truck tow his Tesla away.

But what was mostly on his mind was that companies utilizing AI needed to consider the new ethical and safety challenges. And Horvitz isn't alone. New think tanks, industry groups, research institutes, and philanthropic organizations have emerged, all concerned with setting ethical boundaries around AI.[10]

The automotive sector is rife with ethical conundrums. Consider these (and realize they are only the tip of the iceberg): When programming how a car should respond to an impending crash, whose interests should most be considered—the driver of the car, the driver of the approaching car, or the insurance companies of either? Should the AI of an automobile be programmed to minimize loss of life, even if that means sacrificing its own driver to save multiple lives in another vehicle? How does the preservation of private or public property come into play when programming a car to avoid an accident?

As you can see, the ethical situations AI poses go on and on. And this doesn't even begin to consider AI being used in online ad algorithms, the tagging of online photos, and the relatively new field of private drones.

Horvitz realized rather quickly after calling Tesla that the company was much more concerned about liability issues than they were with solving any deep ethical quandaries around their use of AI. "I get that," says Horvitz, whose love for his Tesla didn't diminish. "If I had a nasty rash or problems breathing after taking medication, there'd be a report to the FDA . . . I felt that that kind of thing should or could have been in place."[11]

These are the sorts of questions we will all have to answer: What will the ethical parameters be for corporate use of artificial intelligence, and who will set the standards as we move into this new world?[12]

Of course, these are only examples where AI is being used *legally*; the bigger cybersecurity threats will come from the use of AI algorithms for illegal purposes. For instance, cybercriminals could, with the help of artificial intelligence, analyze the behaviors of certain people, anticipate their next moves, and then attack them when they least expect it.

Current and future generations, born into a technologically complex world, view cybersecurity in completely new ways and tend to have better-developed instincts than previous generations. They seem less at risk of being compromised by traditional cybercrime tactics. Yet, even millennials have shown themselves—and their personal information—to be overly exposed, as has been demonstrated by the latest events of social media companies abusing their data.

Twenty years ago, the question we wrestled with was whether complete interconnectedness in an everything-is-online world would be worth the resulting loss of privacy. Since then, we have answered that question with a resounding yes, connecting anything and everything to the Internet, including ourselves—through our phones and other smart devices—and introducing a brand-new, overarching question:

In a world where everything is connected, how can we help the individual maintain privacy, security, and autonomy?

As recent social media practices have shown, individuals and organizations will sell data, even private data, when large profits hang in the balance. We have to return to a stronger concern for privacy, using the identification of people as a way to protect them against potential abuse or personal data loss. We must set up an ethical hedge that will keep AI from illegally exploiting those who engage with it. This brings us to the major questions we need to answer:

1. How will an ethical security hedge be created to keep AI from exploiting us?
2. Who should be responsible for this endeavor?
3. Is this a government mandate?
4. Or is the effort left to the private sector?

These questions must be addressed if today's cybersecurity is going to provide us with a legitimate sense of safety and comfort, protect our basic human values and rights, and enable us to thrive in an online, hyperconnected world. Their answers will likely be found in a collaboration between public, private, and government sectors. Are we willing to begin that collaboration now and continue it until we have a satisfactory solution?

6

HEALTH

The sad reality is that at least 3 billion people on this planet have never seen a doctor, don't have access to one, or cannot afford one. It feels criminal that they cannot be cured of their diseases and illnesses, especially if medication or treatment exists.

The future of health care will do much to remedy this. We are on the cusp of medical advancements that we previously believed to be impossible: CRISPR, a powerful tool for editing human genes, is becoming increasingly affordable; cancer treatments are becoming more and more individualized; and diseases that have plagued humanity for centuries are being completely eradicated. In the very near future, the main question surrounding technology and health will no longer be "What can we do?" but "What should we do?"

In one story from Greek mythology, the goddess Eos, who was the personification of the dawn, fell deeply in love with a man named Tithonus. Unable to contemplate a future where her mortal lover would die, Eos asked Zeus to grant Tithonus eternal life, and Zeus consented.

But things did not work out as well as Eos had hoped. She had forgotten to ask Zeus to also give Tithonus eternal youth. While he lived forever, he grew old and decrepit, gradually wasting away—but never completely. He spent the remainder of his days wishing for death.[1]

We are now on a technological path moving toward increased health and longer life, toward greater efficiency in health care, and toward the eradication of disease. But it is also a path paved with ethical questions and technological challenges—chief among them, finding a balance between the length of a life and the *quality* of the life being lived.

> *The woods decay, the woods decay and fall,*
> *The vapours weep their burthen to the ground,*
> *Man comes and tills the field and lies beneath,*
> *And after many a summer dies the swan.*
> *Me only cruel immortality*
> *Consumes; I wither slowly in thine arms.*[2]

"TITHONUS," BY ALFRED, LORD TENNYSON

The Future of Health Is Nearly Here

Traditional nutrition tells us to eat from the food pyramid and exercise. Traditional medicine tells us to treat patients using our knowledge of human anatomy, physiology, psychology, and neurology. Both pillars of human health have long been based on a general approach to specific ailments. Is that really the best we can do? Fortunately, technology has accelerated our efforts to establish a new scientific approach: It's known as nutrigenomics—and it could change how we pursue health care forever.

Nutrigenomics, as its name suggests, is the science of the relationship between nutrition and the human genome. According to the U.S.

National Institute of Environmental Health Science, while genes are critical for determining function, "nutrition modifies the extent to which different genes are expressed and thereby modulates whether individuals attain the potential established by their genetic background."[3] M. Nathaniel Mead, of *Environmental Health Perspectives*, explains that—

> More recently, this [understanding] has been broadened to encompass nutritional factors that protect the genome from damage . . . Researchers recognize that only a portion of the population will respond positively to specific nutritional interventions, while others will be unresponsive, and still other [sic] could even be adversely affected.[4]

What this means is that there is no such thing as a general approach to health care that works effectively for everyone. To this point, explains Mead, "[T]oday's biologists concede that neither nature nor nurture alone can explain the molecular processes that ultimately govern human health. The presence of a particular gene or mutation in most cases merely connotes a predisposition to a particular disease process. Whether that genetic potential will eventually manifest as a disease depends on a complex interplay between the human genome and environmental and behavioral factors."[5]

The application of the individualized health care approach is in its beginning stages. Today, our efforts are focused on understanding Mead's complex interplay; in the near future, they will shift to providing optimal health through personalized nutrigenomic medicine. And once we have aggregated millions of individual results, we can hone in on better and better diagnoses, prescriptions, and lifestyle recommendations. While we push to accomplish this, we should also aim to guarantee the security of every individual's nutrigenomic information across the supply chain.

We are still seeing the early benefits of individualized medicine. One of the major consequences of improving people's health is the stabilization—and even reduction—of the world's population. This has already occurred in developed countries, where the number of children born decreases as health conditions improve. Fewer births are required when birth mortality rates drop and life expectancy increases. Recent analysis suggests that as this Fourth Industrial Revolution spreads, and its improvements make their way around the world, the same will happen in developing countries.

The world's population has reached 7.5 billion people, and we expect that to reach 9 billion by the year 2050—but from there, it's entirely possible for the population to stabilize. This kind of reduction in birth rates would produce massive improvements in other areas of human concern, including food security and pollution.

Another critical focus of future health is the interconnectivity of all aspects of the medical field. Out of the hundreds of billions of objects to be added to the Internet worldwide, estimates are that approximately 20 percent of them will be in the health care industry. By connecting all of their objects, patients, and doctors, hospitals will advance drastically. Ambulances will be able to upload data while in transport—information like a patient's vitals and medical history, the specialties of the paramedics on board, and even their ETA to the hospital. This real-time data will enable more timely and precise treatments when the patient arrives. Not only that, but AI algorithms could begin making life-or-death decisions while a patient is en route, communicating important prognoses to doctors, nurses, and administrators before arrival. That's not to say that medical professionals will no longer be necessary—only that the work these professionals do will look very different. In the near future, becoming a doctor will include far more technological training than it does today.

The Future of the Doctor

People will still visit the doctor—but with a completely different approach. Patients will have greater access to health care at home, including regular check-ups, real-time test results, and cloud-based diagnoses (informed by a doctor's recommendation and their medical records). Those clouds will be maintained and protected using the personally identifiable information of the patient, who will be able to share his or her records with medical professionals all over the world—enabling the right professional to offer advice and intervene, if needed.

As you might imagine, this will dramatically alter the training medical professionals undergo, especially doctors. Where most of today's aspiring doctors study general medicine, in the future, they will be trained to use artificial intelligence and cloud technology to access the aggregated knowledge of the entire medical field.

The doctor's role will also change—likely with a great deal of push-back as it does. Medicine is an extremely wealthy business, and those at the top of the pecking order are not likely to embrace change if it alters the financials, whether personally or corporately. Nevertheless, artificial intelligence can—and probably will—take over many of the responsibilities currently assumed by human doctors. Robots will support more operational procedures and, in some cases, conduct them on their own.

These applications of technology aren't as dangerous as they might sound. Consider that AI and interconnectivity, taken together, will enable doctors to use robotic tools to perform operations from thousands of miles away. This will change—and save—the lives of people who cannot travel to receive a required procedure.

The Future of the Diagnosis

As the use of nutrigenomics expands, the emphasis in medicine will shift from treatment to early diagnosis, with preemptive prescriptions to maximize our health and avoid the sicknesses our genes may suggest. While, up to this point, it hasn't seemed ludicrous to us that 7.5 billion people receive the same prescriptions for pain and inflammation, the growing body of patient data will fundamentally change how doctors prescribe—or don't prescribe—certain drugs. Deep studies into the interactivity of drugs with particular genetic makeups and neurological conditions will ensure that prescribing the wrong medications becomes a thing of the past.

How will the massive pharmaceutical companies react to this? The answer is difficult to guess—and not just from an ethical standpoint. Who's to say that more personalized, molecularized check-ups (paired with the more detailed understanding we'll gain about how our bodies and minds interact with specific drugs, foods, and environments) won't greatly *increase* the number of prescriptions we take. At the least, until humans become a healthier species overall, it's not a stretch to imagine that we will need months (or perhaps years) to reverse the damage we've already done to ourselves through an incomplete understanding of our health.

Our approach to treating heart disease is a good example of the precarious, often-shortsighted place medicine has remained. Heart disease is the number one killer in many countries, including the United States. Yet, most often, the primary treatment options are either bypass surgery or a

74

stent. Both are invasive procedures, bypass being one of the most invasive surgeries a person can undergo. What's so troubling about this approach (notwithstanding the fact that it's symptomatic) is that the American Heart Association uses "class A" evidence—evidence with multiple randomized data points—for only about 11 percent of their cardiac recommendations. On the flipside, 45 percent of their recommendations are based on class C evidence, which is merely expert opinion.[6]

While medicine has progressed light years from where it was a century ago, as we think through how to best approach our health in the future, there will be a plethora of opportunities to utilize the deep mining of personal data—and the aggregation of the world's data through AI—to progress light years further in the next decade. Why wait another century to set ourselves up for longer, more youthful lives? Already, companies are using AI to read an EKG and diagnose an irregular heartbeat (AliveCor)[7]; and to revolutionize X-rays, MRIs, and CT scans (Zebra).[8] Two Pore Guys is building a handheld digital DNA test that's as accurate as lab equipment but far easier and less expensive.[9] Neurotrack is using advanced technology to enable doctors to detect Alzheimer's disease before the symptoms appear.[10]

When it comes to longevity, we've been living our entire lives largely in the dark. This will change dramatically if we are willing to create and embrace technologies that are more adept at performing real-time, molecularized, personalized diagnoses and treatments. There are major (and, we imagine, heated) questions this shift will raise—and is already raising. Some of those surround the notion of equality.

As human life is further extended through a more thorough understanding of our individual bodies and minds, the technologies that enable this pursuit will be first available to developed nations. It is critical that we don't let these advancements begin and end there. We must remember that there are still countries on the planet where malaria and AIDS are taking many lives every year. While money will give certain people access to new developments, perhaps some developments shouldn't only be available commercially. Of course, someone has to pay for the building of these technologies—but perhaps some of them are so important to the health of humanity that the profit models need to shift as well.

The good news, above all other considerations, is that the power of better health will increasingly be placed in the hands of the individual. As this power is transferred, groups of individuals will be both inspired and empowered to share the benefits. But will we? And if so, how? Will it be possible as drug companies drive up the costs of necessary medicines, medical conglomerates drive up the costs of hospital services, and insurance providers drive up premiums?

The future of global health will, in part, be determined along this line of inquiry. In addition, the following questions will help guide the decisions about which technologies should take priority and how quickly the medical profession should adapt to the changes occurring:

1. How will we equalize the benefits of new and improved health resources that future medical innovations provide?
2. Who will take up the mantle of personalized health so that we, as individual patients, have more power, sooner, to improve our lives?
3. How much will one's personal wealth determine the kind of treatment received?
4. How should we navigate the prioritization of longevity against the quality of life? Is longevity the ultimate goal, or is it the highest-quality life, up to a certain age? If the latter, what age is appropriate?

5. As AI takes on a larger role in our health care, how quickly should we trust it?

6. How could we ensure that our health data is protected and not revisable (to the point that a certain diagnosis could be life-threatening)?

7. With regard to birth, treatment, and death, whose desires will take precedence: the patient's, the patient's family's, or the state's (or some other ruling body's)?

We are on the cusp of redefining what is possible when it comes to global health. But we need to know what we're ultimately after before we get there. The more united we are, the more progress we can make. If we remain fractured, a dangerous detachment is likely.

7

JOBS

In chapter 1, we heard Susan Fowler's story about joining Uber—and it left us wondering how Uber's drivers could keep going. We might have even felt anger boiling up at the injustice—and perhaps we were justified. But the fact remains that Uber is not the only one to blame. We're now living in what most call the gig economy: ripe with the overarching promise of entrepreneurial freedom, with no bosses, no set hours, and no offices required. It's not just being promoted by Uber-like companies, who do provide a more flexible way to make a living—it's being commoditized everywhere we turn, even by companies and products who can offer no more freedom than a fancy bottle of water, coffee, or alcohol might provide.

On the one hand, perhaps companies like Uber, Lyft, and Instacart (which Fowler points out are valued at $72 billion, $11 billion, and $4 billion, respectively) are merely cashing in on a sweeping human desire playing out within workforces across the globe.[1] On the other hand, how dare they utilize technology to exploit a healthy human desire for career freedom! This dichotomy characterizes one of the main challenges we face in applying new technologies to human labor.

In the plus column, technology offers us hope that we can experience life outside the confines of a carpeted cubicle without negative consequences. (Perhaps we'll go watch our child's soccer game or take that long lunch with a visiting friend.) Today, we have new career options—options we previously couldn't pursue without taking huge financial risks. Today, we can gig our way to the lifestyle we desire.

In the minus column, technology threatens to upend the stability of the jobs that do exist. If you're someone who already enjoys your work and the flexibility it provides, technology is an antagonist. If you're somebody who needs the stability of the job you have to finance your "test run" in another field, technology is both an antagonist and a protagonist: It gives and it takes away.

Robots are projected to create 15 million new jobs by 2027—and could eliminate 25 million jobs during the same time period. An incredible 38 percent of jobs are at "high risk" of replacement by robots and AI in the next 15 years.[2] So, perhaps we ought to be both thankful and uncertain about the gig economy.

One fact remains: The entire concept of employment is experiencing disruption at a record speed and on multiple levels. It is our job—the one job we all hold, if you must—to ensure that this transition is handled wisely in regard to the capacity and role technology will be allowed to play. In other words, we must share the sentiment of the first coworker who approached Susan Fowler, as opposed to the giggling programmers who seemed to think that arming technology with the power to hamstring human labor isn't a terribly detrimental idea.

As it is, AI is already sweeping through many industries, looking for opportunities to improve efficiency, productivity, and even innovation. In Amazon's massive web of warehouses, many of the packages are now picked from the shelves and moved to trucks by robots. Between the 2015 and 2016 holiday seasons, the expansion of Amazon's robot workforce increased by an astounding 50 percent, to around 45,000 robots in 20 fulfillment centers. Not only did Amazon save on employment costs—their efficiency increased. A task that may have taken an hour or more for a human to complete is now finished in about 15 minutes.[3]

Robots could also replace 25 percent of all U.S. combat soldiers by 2030, according to recent statements given by U.S. Army Gen. Robert Cone. These robots could be used for anything from dismantling land mines to manning the front lines, decreasing human casualties and lowering costs. Being a soldier may become a thing of the past.[4]

Even for jobs you wouldn't expect a robot could accomplish, employers are looking into automation. The next time you board a cruise ship, for example, your bartender could be a robot, one that can make two drinks every minute, or up to 1,000 drinks every day. It knows every drink in the bartender lexicon and produces up to 65 of them before needing its reservoirs refilled. Best of all, this sort of employee will never wake up hungover or negotiate for higher wages.[5]

Robots are filling more and more jobs across every sector. The acceleration of this phenomenon will increase as the Fourth Industrial Revolution emerges.

The Real Question: What Will Jobs Become?

Robots aren't only supplanting human jobs. They are leading us down a path where the definition of the word "job" is changing.

This widespread use of AI and robotic automation is going to change the fundamental ways that we think about work and employment. The concept of "having a job" as a way of maintaining an income—enabling you to live a normal life, pay your tuition, your bills, or your mortgage—will become obsolete as permanent jobs disappear from the landscape. A constantly fluctuating market, along with the short life cycles of many emerging companies, will lead to the near-extinction of full-time jobs—or of an individual's services being sold from one company to the next. Many people, along with the limited skills they possess, will simply become redundant. Having a job will not come with the same stability or guarantees it had during the last 60 years.

What will jobs look like in the Fourth Industrial Revolution, then? First of all, they will almost always be defined by the fulfillment of short, defined tasks. Employment as we know it will be broken down into individual assignments. People with a specialized gift, skill set, or field of knowledge will be able to sell their services to the highest bidder, chosen from a plethora of potential buyers on the Internet. This is how a worker in the Fourth Industrial Revolution will create a constant income flow, hoping to adequately subsidize their lifestyle.

The good news for workers is that this molecularization of job activities will lead to almost complete autonomy. Individuals will not have a boss to report to, a company overseeing them, or an office where they have to check in every morning. Each worker will become his or her own company, attempting to endlessly capitalize on their skills and expertise in a marketplace made up entirely of online customers. Yet, this will also lead to worldwide competition for work, because people located anywhere—using a computer or mobile device—will be able to broker their services throughout the global community.

More and more ancillary responsibilities will be assigned to robots and personal assistants—but not only by large corporations that are looking to increase workplace efficiency. Individuals will also utilize robots to increase the productivity of their own mini-corporations. AI will clean the house or order groceries, freeing up more time for the individual to focus on their primary, income-producing work. Robots will even replace the phone responsibilities of executive assistants and bank tellers, jobs that have complex interactions with customers, and in many of these cases, the customer will not be able to detect they are not talking to another human.

All of these changes will accelerate as humans become more inclined to devote their time to productive activities. Of course, many new, unforeseen jobs will be created in the Fourth Industrial Revolution.

The ways people are compensated for their work will also evolve. They will be paid in cryptocurrency, stored in credit- or debit-based crypto-wallets associated with their digital identities (depending on the type of transaction). For instance, someone may be able to sell parts of a task to one person or organization and, at the same time, subcontract other components of that task to a different person, who will be connected to the same financial and task ecosystems. Using cryptocurrency will decrease the friction in these types of transactions and eliminate (or at least reduce) the need for a bank account—lowering fees for everyone and increasing efficiencies.

In some of the wealthier developed countries, more people (think baby boomers) are reaching retirement age than ever before, exponentially increasing the number of jobs focused around the care of the aged. These new jobs must be filled by caretakers: people who display certain skills, including the ability to treat others with empathy, charity, and love.[6]

Robots, in some cases, may never be able to entirely replace human interaction.

Education in a World Without Jobs

So, if the concept of having a job is reframed in this new industrial revolution, what happens to groups like universities or trade schools, organizations whose primary function is to prepare individuals for a particular workforce? If there are no jobs in a given field, there is no need for a diploma to verify one's training. This shift will deconstruct the entire education system, in both developed and developing countries.

Education, like almost everything else, will increasingly be conducted via the Internet (and other new communication tools)—in a provided-as-needed fashion. In order for people to improve their market offerings, they will still need to develop certain skills. Any training required will be tailored to them individually, in specific courses or tracks of learning, almost in the same way you receive a prescription. It will be obtained directly, over days, weeks, or months—not as part of an extended, multifaceted four-year process.

This molecularization of education is of the same type we've discussed for health care and employment. Everything about the Fourth Industrial Revolution points to specificity and the provision of goods and services in a highly individualized manner, giving rise to entirely new industries (or, at least, new subindustries within major ones).

We are already observing new jobs crop up in areas such as robotics manufacturing and drone operations. There are also peripheral jobs developing because of these same advances, including solar panel installers and cryptocurrency specialists. These jobs did not exist even five years ago.

Factored together, all of these issues bring us to the conclusion that, in the next century, many humans will work in a way that is more focused and specialized than ever before. This is becoming the case in many Middle-Eastern countries, where the resource-to-population ratio is high and almost all of the manual labor is being delegated to immigrants. In the very developed countries, such as Switzerland, it is rare to see the native population performing manual labor. We already find certain tasks being consigned to a particular subset of the population—or even off-loaded to robots, in the case of Amazon.[7]

It is now foreseeable that we will delegate more and more tasks to both digital AI assistants and physical robots—to the point that the concept of the job as we know it today will put on an entirely new face. There are big questions surrounding this shift:

1. How should we utilize the newfound job freedom that the application of AI and robotics provides?

2. Can this new era of jobs (and in some ways, joblessness) be used to bring greater equality to an economically disparate world, or will it only serve to increase the gap between the haves and the have-nots?

3. What other ways can we resource our newfound freedom to provide better and more individualized careers for ourselves?

4. How should society reach out to those whose jobs are being replaced in the near term? Are there opportunities for new careers in this space?

As industries are upended and human jobs are replaced by machines, we must not forget that work has always been at the center of our lives and identities. It will likely always remain so. But perhaps the time is coming for a new definition of work—work defined by tasks that are less mundane, that contain more meaning. Does the horizon show signs of more meaning for us, even if the process might be a little painful?

The truth is that most of us have little time to be identified by anything other than our jobs. We spend the vast majority of our adult lives working, formally employed by someone else (or self-employed). We spend fewer hours at home and identify less and less with our family roles; this is certainly the case in many developed nations. We also have less and less margin to pursue our interests. Furthermore, our weekly relationships are more often defined by our workplaces than, for instance, our hometowns or childhood friends. All of this adds up to our identities being largely tied up in what we do for a living—and the margin we do have, in recovering from our jobs.

To paraphrase Al Gini in *The Importance of Being Lazy*, the unfortunate reality for too many of us is that our various forms of recreation and relaxation are really about recovery and recuperation, rather than rapture and rediscovery. This is because our diversions from work are usually momentary distractions from the pace of life. They've become geared toward overcoming fatigue, numbing awareness, or appeasing a particular appetite—so that we can go back to the job to endure and earn more.[8] "Ultimately," writes Todd Duncan in his classic book *Time Traps*, "when our time is monopolized by our work—and/or recovering from work—the only thing that forms our identity is work. We become lost in our job."[9]

Ultimately, we—not technology—must define our future work, clarifying the indispensable human skills we will require and then putting them into practice within the new economies that unfold. And then, if we so choose, we can allow these efforts to be governed by a commitment to expanding our identities and including all we are as human beings, not just workers.

8

MONEY

As we've indicated throughout this book, modern technology should enhance the lives of humans. It should make the difficult easier, increasing our ability to satisfy our core human needs. In this era, we meet those needs—food, clothing, and housing—through a transactional, market-based economy powered by complex financial institutions. Greater access to those markets leads to an increased potential to fulfill those basic requirements. But as we've seen in the decade following the financial crisis of 2008, modern markets are laden with risk. And what about those who have no access to those markets in the first place? How difficult is it for them to get a leg up in the world? Can modern technology solve these issues?

We believe it can.

Perhaps no sector is currently being disrupted by modern technological advances as much as the financial sector. *Blockchain Revolution* author Don Tapscott recently shared with us how the application of modern technology to the financial sector is revolutionizing the world, and not just for the wealthiest among us.

Referring back to his book, Tapscott told the story of Analie Domingo, a Filipino housecleaner living in Toronto. As an immigrant day laborer, Analie works hard for her money, but she isn't working to support only herself. Like many migrant workers, Analie's income helps support her family—in her case, her mother, living in Manila. But how does Analie send that money? In the early days of living in Toronto, it wasn't all that easy.

When she first arrived in Toronto, Analie used one of the more traditional methods for transferring money. Each week, she took a portion of her wages and made her way to the local Western Union branch. There, she filled out time-consuming paperwork (the same paperwork each week) authorizing Western Union, the financial courier, to transmit the money to her mother. Of course, the transaction came with a cost, and fees amounted to approximately 10 percent of the transferred funds. Leaving Western Union, Analie would then trust it to deliver the money to a branch in Manila, and her mother could make the trip into town to pick up the funds.

The transaction was time consuming, often taking several hours (factoring in the walk to the Western Union branch, the time to fill out the paperwork, and the walk home). It wasn't efficient, either. Often, almost a week would go by before her mother received the money. The transactions were slow and costly, but Analie continued to use Western Union because it was a proven intermediary—an intermediary that had earned her trust. And though the delays and expenses were cumbersome, Analie didn't see another method for transferring funds home. That is, until she discovered Abra.

Abra is a cryptocurrency wallet that allows low-cost peer-to-peer transactions. Using Abra, Analie began purchasing cryptocurrency, digital currency that's supported by blockchain technology. That currency was issued directly to Analie, stamped with her unique, nearly unhackable digital fingerprint, and recorded on a dispersed and encrypted global ledger (which stores the particular transaction history of the currency units she purchased). Once the cryptocurrency was in her Abra wallet, she'd send it to her mother's Abra wallet. Within seconds of the transfer, her mother received a notification of receipt and could then locate an Abra teller through the smartphone app. Once she found one within close proximity, she'd schedule a delivery, and the teller would arrive within minutes. By pressing a few buttons, she transferred the cryptocurrency to the teller, who gave her Filipino pesos in exchange, taking only a 2 percent transaction fee. And because of the blockchain encryption associated with the transaction, the whole thing occurred without the exchange of personal data and without risk of theft.[1]

By eliminating the intermediary and using a peer-to-peer method of exchange instead, Analie saves time each week. Her mother receives funds much faster and retains more of them. And perhaps best of all, the transaction occurs with less exchange of personal data. It's all possible because of the underlying technology known as blockchain.

Blockchain technology has connected people to financial services like never before, giving them new access to capital markets and methods of exchange. Could this be the answer for many of the global financial challenges we face, particularly for those people with less access to capital?

The Financial System: A System of Trust

Our capital markets and economic systems have developed over thousands of years, and over those years, one fundamental tenet has been necessary to ensure they operate properly: trust. In modern developed societies, economic trust is often placed in large intermediaries—including banks and other financial institutions, governments, and even social media companies (e.g., Facebook and Venmo). Those intermediaries authenticate parties, settle transactions, and keep transactional records. But over the years, trust in the financial systems has eroded—due to the bad acts of both individuals and the institutions themselves.

Historically, society's institutions jealously guarded the trust the public placed in them. They took care of our money, recorded our transactions, and refrained from risky investment strategies. But in the run-up to the 2007 financial crisis, things shifted. American banks overextended themselves on loans to subprime borrowers—borrowers without the financial resources or collateral to pay their loans. These loans were encouraged, some say, by the U.S. government's deregulation of the financial industry. Those subprime loans were packaged together in tranches of securities and sold to institutional investors. Those securities were then insured by credit default swaps, a risky form of insurance issued by other large financial institutions. But as property values plummeted and subprime consumers defaulted on their loans, the worth of those securities cratered. The losses were so deep, in fact, that the insurance companies couldn't cover them. The result? The issuers of those credit default swaps went belly up, institutional investors lost billions, and too many consumers lost their homes.

The losses as a result of this credit crisis nearly collapsed the global economy. It was a collapse only staved off by the U.S. government, which—after having deregulated the industry—poured billions of dollars into the financial institutions that created the problem in the first place. In essence, two of our largest institutions had breached our trust.

The financial collapse alone would be reason enough for the erosion of the public's trust, but the breaches of trust keep piling up. In recent years, our institutions and financial intermediaries have been the subject of multiple cyberattacks. In 2014, JPMorgan Chase suffered a data breach, allegedly conducted by Russian hackers. In 2017, Equifax announced a cybersecurity breach that affected over 145 million United States consumers; hackers gained access to users' full names, social security numbers, credit card information, birth dates, and home addresses. In 2018, it was discovered that (again, through a data breach) 87 million Facebook users had their personal data shared with Cambridge Analytica.

Over the years, we've placed our trust in these companies, but at the end of the day, that trust has been breached—often as the result of attempted innovation or technological failure. And though we're no experts in breaches of trust resulting from poor lending practices, as experts in the field of cybersecurity, it's our opinion that many financial institutions aren't equipped to protect our data. Much of the server-side security technology employed by those institutions is outdated or under-powered. So, as long as cyberhackers have a motive to obtain personal and financial data, those institutions and corporations will be subject to potential data breaches and hacks.

Of course, we can invest in new technologies to try to prevent cyberattacks—and even provide analytical data to help us avoid the next financial crisis. But investing in these strategies requires that we place even more trust in institutions that have already eroded our reasons to trust. And investment in those areas of our economy does nothing to increase global prosperity or create financial security and stability for the masses without access to capital markets.

As of the writing of this book, there are over 2 billion people without access to the modern financial system. They don't have a safe place to stash their savings and cannot transfer money to friends and family members outside of their community. They have no access to home or commercial loans. And because the global economic system doesn't work for those on the outside, people like Analie Domingo's mother are often taken advantage of. Without access to modern banking institutions, transaction costs for transferring money can range between 10 and 20 percent.[2] What's more, because institutions such as trust companies, title insurance companies, banks, and (stable) governments work together to record, insure, and protect ownership of real estate holdings, their greatest stores of value—their homes and farmlands—are less than secure. Some believe that, in the developing world, up to 70 percent of landowners hold a flimsy title. Often, those titles are stored on a government computer (or in a filing cabinet) and are subject to the whims of a dictator, who might order the land records changed so that he can take ownership of the property or gift it to his friend.

In short, those within the modern economic system are forced to place ever-increasing amounts of trust in those systems (flimsy as they may be from time to time), while those outside the system operate at a significant disadvantage. All of this undermines the basic human need for economic security. And though this may seem like an untenable economic situation, it's indicative of a significant opportunity. What if we built a new financial system, one that does not depend on the public placing trust in vulnerable institutions or corrupt and bloated bureaucracy? What if we applied technologies in ways that helped us create, transfer, and preserve wealth, all at a relatively low risk? How do we do that? Here enters the concept of blockchain-based technology.

The Rise of Blockchain Technology and Cryptocurrency

In 2008, a computer scientist (using the name Satoshi Nakamoto) wrote a whitepaper entitled "Bitcoin: A Peer-to-Peer Electronic Cash System."

It was the year after the financial markets melted down, and many were looking for ways to decentralize capital and create more universal mediums of exchange. In the whitepaper, Nakamoto showed how commerce on the Internet had "come to rely almost exclusively on financial institutions serving as trusted third parties to process electronic payments."3 And though that system worked well enough for most transactions, Nakamoto argued that it suffered from the inherent weakness of any trust-based model. Nonreversible transactions are impossible, because there will always be disputes that need mediating. The intermediary role of the institutions in settling those disputes, in ensuring funds were moved from buyer to seller as authorized, increases transaction costs, too, which makes smaller, casual transactions financially infeasible. And with the possibility of transaction disputes and reversals, merchants need to be more wary of their customers. They need to acquire more data than they would when money changes hands in a face-to-face, in-store transaction.

These things drove the cost of transactions up, Nakamoto argued, and placed too much power in the hands of these intermediaries, most of which were financial institutions. He said that if the role of these intermediaries could be limited, if a type of currency could be created that relied on cryptographic proof of authenticity instead of trust, a more accessible and affordable peer-to-peer monetary system could be created. There was one problem, though. In a peer-to-peer payment system, how could a recipient of funds ensure that the transferred money was a true representation of value? How could they ensure the money wasn't a digital copy created by a fraudster or hacker? In short, how could the payee avoid the "double spend" problem?4

Nakamoto proposed an elegant solution: a digital currency with a chain of digital signatures. That chain would be kept on a dispersed and encrypted public ledger, and each user would have a unique digital signature, applied to the digital currency so that any payee could verify the coin actually belonged to the holder. This public key, or public ledger, would trace every use of every coin to ensure there were no attempts to copy or "double-spend" the currency.

With that paper, the idea of Bitcoin—a blockchain-based currency— was born.[5]

The Blockchain as a Means of Meeting Human Needs

Bitcoin, Ethereum, and a variety of other cryptocurrencies have gained cult followings since Nakamoto's paper was first published. Those currencies have had their own issues, including wild fluctuations in value. For instance, as of January 1, 2017, the price of one Bitcoin was $963. By January 6, 2018, the price of a Bitcoin had risen to $17,712. By the end of August of that same year, the value had plummeted from its high, and the price of one Bitcoin was only $6,891. To make matters worse, in June of 2018, news outlets reported that a South Korean cryptocurrency exchange was hacked, leading to the loss in approximately $40 million worth of currency. And though that was a relatively small amount of currency, the hack raised questions about the security of cryptocurrencies in general. As a result, the cryptocurrency market shed billions of dollars of value.

Despite the volatility and the rare, but present, security risks, Bitcoin and other cryptocurrencies have delivered on their promise. They've cut intermediaries and large institutions out of many transactions and have allowed consumers to control their personal information as they engage in peer-to-peer transactions. What's more, they've created an alternative financial system with relatively few barriers to entry. Anyone with a computer (or smartphone) and an Internet connection can exchange cryptocurrency, often with only nominal transactional costs. Simply put, cryptocurrency connects those outside of institutional economic systems with modern markets, all the while minimizing their exposure to theft and fraud.

Approaching $100 billion globally in 2018, peer-to-peer financial transactions are growing at a feverish pace in China, where today smartphone users top 1.6 billion. Dr. Wang Wei, considered the pioneer of Western-style investment banking in China, is perhaps the country's strongest advocate for financial evolution, having created nine finance museums across China and countless publications to inform, educate, and connect Chinese citizens in the changing face of finance. "Right now 60 percent of the population of China are young people under the age of forty, who grew up with the internet. This group is very familiar with Bitcoin and other cryptocurrencies," Wang Wei informs us. In 2015, Dr. Wang helped establish the Institute for China Blockchain Applications. With more than 2000 members and 50 institution directors, the group now has eight local centers, four in China and four abroad. Wang Wei adds, "I always tell young people, that while you cannot use cryptocurrency in our country, with your talent, you can use the blockchain theory to do many other things. And the applications for blockchain will reach far beyond cryptocurrency. I really have confidence that China will be a leader in this blockchain revolution."

The use of cryptocurrencies as a way of providing instant access to a more modern financial system is fascinating, but the underlying technological concept—the blockchain—has implications that reach far beyond money and markets. The blockchain, a cryptographic technology that applies individual digital fingerprints to both people and objects (like cryptocurrency)—and keeps a public ledger of transactions associated with those individuals and objects—could go a long way to solving many of the world's property challenges. For instance, if deeds and titles to real property were no longer held on paper, if they were digitized and included blockchain technology, land could be transferred peer-to-peer all over the world, without the need to involve trust companies, governments, or other financial institutions to ensure the chain of title. What's more, those deeds and titles would be stamped with a unique, digitally encrypted identification associated with the owner's own digitally encrypted ID. And because those unique identifications would be independently verifiable—if held on third-party servers outside government control—those titles would be virtually hackproof and not subject to government theft or fraud. In this way, the blockchain could provide security for those whose largest store of value is held in their property, thus meeting a fundamental human need.

The application of blockchain technology won't be limited to finance and real property, though. It'll be applied through the financial system in ways that make our lives easier. Imagine a future in which your connected refrigerator (as a blockchain participant) exchanges your cryptocurrency (authorized through your blockchain) to order milk (tracked on a blockchain supply chain system), delivered by a drone (as a blockchain intermediary). Through the blockchain, transactions will be more seamless, more fluid, and will take less of our time. It's an elegant solution for a more connected age.

The Biggest Advancement Since the Internet

The blockchain has the potential to be the biggest technological innovation since the advent of the Internet, and it might cure many of the problems endemic to our modern financial systems—problems like breaches of trust, losses of data due to hacking, and losses of value brought on by institutional greed or government corruption. By pushing out middlemen and intermediaries, the blockchain might give consumers the power of control over their data. And by decentralizing financial transactions, it minimizes our reliance on large financial institutions, it makes transactions cheaper and more fluid, and it allows for microfinancing opportunities that are otherwise cost prohibitive. What's more, through the unique, distributed, universal ledger, it brings accountability to financial institutions and governments.

If we're committed to using modern technology in ways that enhance our lives, that help us attain our basic needs and wants, we need to take a hard look at our financial systems. We need to ask whether there are ways to decentralize it, to give the power back to the people. And in that, we need to ask the following kinds of questions:

1. In what ways could we use technologies like the blockchain to make transfers of capital easier, thus allowing for a more equitable distribution of wealth?

2. How can we explore and advance blockchain technologies that protect land ownership from theft by fraud or corruption?

3. How might we use decentralized systems of currency to give small business owners in developing countries the access they need to capital for the expansion and scaling of their businesses?

4. What companies are pioneering the kinds of technologies we'll need to ensure a secure blockchain, and how can we invest in and partner with those companies?

Examining how the application of technology might advance the best interests of humans with respect to financial security, one solution stands out among many others: the blockchain. As we advance this technology, we find ourselves poised on the edge of a great opportunity, the opportunity to level the economic playing field for billions of people across the globe. It could change the economy as we know it, and if used properly, it could serve as a method of wealth repatriation (instead of wealth redistribution). It could level the playing field for so many who are on the outside, looking in—folks like Analie Domingo and her mother.

9

TRANSPORTATION

In the beginning, humans looked for more efficient ways to get from point A to point B. We walked. Then ran. Then climbed atop a horse. Reaching the limitation of horseback transportation, we innovated. We created the internal combustion engine and put it into boats; then, with a few modifications, automobiles; and eventually, with more modifications, airplanes and rockets. And because those engines burned up fossil fuels and polluted the air, we innovated again, this time creating the electric engine. The story of human transportation is a story of constant innovation, constant disruption. But is this constant change always a net positive?

In 2018, San Francisco was overrun with Birds—Bird scooters, that is. There were Birds everywhere. The sidewalks. The streets. Under the awnings of any given storefront. The people of the Golden Gate City weren't happy about it. Birds were making a mess of things, they said. They were noisy and didn't respect the average pedestrian's personal space. And the "sudden surge," as one local news anchor put it, had come without any warning.

It'd have been bad enough if it had only been the Birds, but it wasn't.

Limes littered the streets. Spins too. Three companies—Bird, Lime-S, and Spin—had dropped their electric scooters onto the streets of San Francisco without permits or permission, and the influx of this eco-friendly mode of personal transportation was the source of much discussion—most of it negative.

One news outlet reported that a constituent emailed a city supervisor complaining that he'd broken his toe when he tripped over a scooter left in the sidewalk.[1] Others complained that the scooters were zipping in and out of traffic, causing problems. Some claimed scooter operators were endangering slower-moving pedestrians on the sidewalk:

> "I almost got hit three times."
>
> "Obviously they don't belong on the sidewalks, but that's where they are."
>
> "Get rid of them!"[2]

There were those who loved the mode of transportation, of course, but the detractors were vociferous in their objections. And it wasn't just in San Francisco, either. In Santa Monica—a suburb in Bird's backyard—city officials alleged that Bird dumped their products onto the street without proper permits. And though Bird claimed they were operating in a gray area of transportation and business regulation and didn't need any special permits to lease their electronic scooters to the public, city officials disagreed. They disagreed so vehemently, in fact, that the Santa Monica City Attorney's Office filed a criminal complaint against Bird, alleging the transportation start-up had violated no less than nine criminal statutes. Ultimately, Bird capitulated, agreeing to pay over $300,000 in fines to the city—and further agreed to seek the proper permits.[3]

Travis VanderZanden, founder of Bird and the former COO of ride-share company Lyft, is well aware of the growing pains that come with being a disruptive force in the transportation industry. He hasn't let these growing pains hold Bird back, though. In a 2018 interview with CNET, he said, "Our mission is really to help reduce car trips, traffic, and carbon emissions . . . We think Bird is having a very positive impact in the cities we're operating in."[4]

But though electric scooters might reduce traffic and carbon emissions, though they may help people get from point A to point B more efficiently, are they inherently good? Do the negative effects (indicated by the pedestrian outcry) outweigh the positives? Did the electric scooter companies advance a technology in markets that were unprepared? Did they give the local authorities enough time to regulate and minimize the disruption to the flow of city traffic?

In this Fourth Industrial Revolution, advances in technology are disrupting methods and modes of transportation faster than ever. But if we're to adopt those disruptions, we'll need to ask some hard questions, including this one: Are we prepared to be disrupted?

Transportation in the Fourth Industrial Revolution

Historically, transportation has been among the first industries to benefit from the application of new technologies. With the advent of the steam engine in the First Industrial Revolution, people began to travel more and transport goods and services over greater distances.

In the Second Industrial Revolution, advances in railroad technology led to the construction of transcontinental railroads, allowing people access to previously unreachable territories. And with the advancement of cars, such as Henry Ford's Model T, humans enjoyed a level of autonomy in transportation that was previously unknown. For the first time in history, we could cover great distances at relatively low costs, all while staying connected with the world beyond our own communities.

The Third Industrial Revolution brought the aviation industry, which opened up global travel. No location on the globe was off limits, and we could get to even the most remote destinations in a relatively short amount of time—at relatively accessible price points.

But technological advancements in transportation haven't come without a cost. Because advances in transportation have risen on the back of the internal combustion engine, have prioritized personal autonomy, and have connected us in ways previously unknown, the application of technology to the travel industry has created so many unintended consequences. The emission of greenhouse gases; congestion caused by so many travelers; the cumbersome and sprawling transportation infrastructure required to sustain planes, trains, and automobiles; the opportunities for international terrorism—all of these things have resulted from advancements in transportation. And though these are generally acceptable drawbacks when weighed against the positive aspects of travel, technology can ameliorate those risks in the Fourth Industrial Revolution.

The travel industry is poised to be one of the most technologically disrupted fields in the next 20 years—so disrupted, in fact, that the modes of transportation you use in a decade may more closely resemble those used by your favorite science-fiction characters than those you use today.

Imagine with us. You're walking quickly along a busy city sidewalk, jostled by the crowds of people walking in the opposite direction. You glance at your watch without stopping and groan because no matter how fast you walk, you're not going to make it to your meeting on time.

You're too far from a subway station, and the streets are gridlocked, ruling out an Uber, a Lyft, or a traditional taxi. The buses are barely moving in the traffic.

What can you do?

Hail a flying taxi, of course. You pull out your cell phone, and with an account much like Uber or Lyft, you request a vehicle, but not one affected by silly little things like ground traffic and congestion. A taxi comes from the sky, squeezes into an open spot next to the curb, and you crawl in. The driverless, drone-like vehicle whisks you off to your meeting, and you arrive on time.

No, you're not on the set of *Blade Runner 2049*. You're in the reality created by Rolls-Royce, a reality scheduled to make its appearance in the mid-2020s. In the summer of 2018, just before the Farnborough Airshow, Rob Watson, director of Electrical at Rolls-Royce, revealed that the company is working on an electric vertical-takeoff-and-landing (eVTOL) vehicle that could carry up to five people. The eVTOL taxi could achieve speeds of 250 miles per hour and travel a distance of 500 miles, all with a much smaller carbon footprint than a car.

All of it is part of Rolls-Royce's desire to get ahead of the game in personal air mobility. "We believe that given the work we are doing today to develop hybrid electric propulsion capabilities, this model could be available by the early-to-mid 2020s, provided that a viable commercial model for its introduction can be created," a Rolls-Royce spokesperson told BBC.[5]

Flying electric taxis won't be the only disruption to modern transportation. As of the writing of this book, communities across the globe are considering employing hyperloop technology for mass transit through mostly underground tunnels. Hyperloop developments in the United

States, Europe, and Japan will slowly replace traditional high-speed trains, which take up enormous swaths of land. With this mode of travel occurring primarily underground, the majority of old railroad lands can be used for other purposes.

In addition to advancements in ground transportation, the application of technology to aviation may alleviate the negative by-products of travel. In 2018, it was reported that aviation start-up Zunum Aero hoped to make regional travel more accessible by using hybrid-electric aircraft, which would cut down on carbon emissions and reduce the overall price of travel by passing fuel savings along to the customer. According to a *Fast Company* article, Zunum "plans to have its planes in the air by 2022." The planes might be piloted by humans, or they may be automated, and the aircraft wings might contain modular batteries to power the plane. In an effort to increase flight distances, gas turbines will power a generator to supplement the battery during flight. And because batteries could simply be swapped at the terminal, there'd be no need for refueling, thus cutting the turnaround time for any flight down to somewhere around 10 minutes.[6]

International transportation will undergo its own sci-fi sort of disruption. Some in the travel industry dream of a future where air travel occurs at much higher speeds (perhaps as much as 16,700 miles per hour) and at much higher altitudes (perhaps just outside of the terrestrial orbit). But these aren't just unfounded dreams. Folks like Elon Musk are working to make it a reality through his company SpaceX, a company that hopes to apply the principles of interplanetary transportation to international travel. And if he's successful, he estimates a trip from New York City to Shanghai could take as little as 39 minutes.[7]

But these disruptions won't come overnight: They're years away. Other disruptions—disruptions like those caused by electric scooters or rideshare companies—will be more immediate. As rideshare providers like Uber grow, and as their usage grows by 5 or even 10 times in any given city, vehicles that once drove 12,000 miles a year might be used to drive as many as 100,000 to 200,000 miles per year. And as this becomes more of a reality, the move away from gasoline-powered vehicles and toward electric cars will be more cost efficient and help reduce carbon emissions. In addition, as these companies grow and technology advances, eliminating the driver altogether will reduce costs and liability risks. In this way, what once constituted the largest business expense—human capital—will all but disappear. And if those cost savings could be passed on to the consumer, the reduction in the price of transportation might make it difficult to justify the expense of owning a personal vehicle, especially in more urban areas, where alternative modes of transportation are readily available.

In the not-so-distant future, smart cities will offer point-to-point rides on demand, dramatically reducing costs for customers and citizens, not to mention time wasted waiting at the bus stop. Those point-to-point rides might be through connected pods, or driverless cars, or smart trollies—all of which would be run by electricity, again leading to reductions in our transportation carbon footprint. In addition, because electric engines are simpler and have fewer moving parts, the rides themselves would be more reliable.[8] In theory, these advancements might lower the total numbers of cars on the road, thus reducing congestion and environmental pressure.

Disruptions to Transportations of Goods

The technological disruptions to transportation will have a dramatic impact on the way people move about the world. Just as importantly, though, they will impact the way goods and services are delivered. Everyday consumer goods—food, clothing, office supplies, and other sundries—will be delivered digitally, or by drone, or by some sort of dedicated hyperloop. And so much of that transportation will be automated.

Consider the refrigerator example we've been using. The fridge detects a shortage of milk and (based on your criteria) it places an order to the local supermarket (using encrypted and personalized communication through an IoT network). It then contracts with a delivery drone (again, by way of an IoT network) to deliver that milk. The drone picks up the carton and delivers it directly to your refrigerator. Your refrigerator pays for the milk (using cryptocurrency per your authorization). Upon payment, an automated system at the warehouse transfers the ownership of the carton to you. The drone returns to its owner, waiting for another order from somewhere else in the city. This automated and connected web of transportation, communication, and finance simplifies supply chain management and reduces the need for human transportation.

As fascinating as changing modes of transportation might be, the most significant advancements may not be seen at all. As transportation becomes smarter and more automated, people, objects, and products will be plugged into an ever-connected worldwide transportation platform. Each person, object, or product will have its own personal and unhackable blockchain identification, and the platform will allow every participant to develop an algorithmic profile that can be updated, maintained, and protected in real time. The data generated by this platform will enable governments, transportation companies, and even family members to analyze personal transportation needs, to respond to those needs more efficiently, and to govern and regulate (and in the case of families, monitor) transportation with more precision and thoughtfulness.

Reinventing Transportation

A transportation revolution is upon us, and with it, nothing is impossible. Driverless technology, electric cars, reimagined mass transit solutions, battery-powered vehicles, dedicated self-driving public transit lanes, and mobile transit scheduling solutions will all be realities in the future. But aside from lower transportation costs and reduced environmental impact, what could the benefits of this sort of transportation revolution be?

Real estate could be liberated from parking lots and replaced by parks or sustainable housing. Faster and more environmentally friendly modes of transportation might allow us to create smaller communities outside urban centers, all without sacrificing access to those centers or the goods and services they provide. Drone delivery, point-to-point transportation, underground hyperloops, and flying taxis might reduce the amount of street-level vehicle traffic exponentially.

Consider the international benefits, too. Faster, more fuel-efficient, and more cost-effective international travel might help us expand our worldviews and might increase the spread of common political ideologies. In communities across the world, empathy might expand as we come into contact with new and different peoples and cultures. Pollution, in places like India (which is home to 9 of the 10 most polluted cities in the world), Pakistan, and China, could be dramatically reduced with the application of electric transportation technologies, and day laborers living hundreds of kilometers outside of bustling cities might finally have access to employment.[9]

The Risks of Modern Transportation

For all the benefits of applying modern technology to transportation, there may be unintended consequences, too. With the ability to travel over greater distances in less time, people may be tempted to live even farther away from places of employment and urban centers. And if this results in the creation of smaller, closer-knit communities, that might be a net positive. But if it leads to greater isolation, to more autonomy and less connection, it might ultimately have negative impacts on society at large.

The automation of transportation might have detrimental effects on those employed in the transportation industry, too. Once modes of transportation become driverless and pilotless, rideshare drivers, deliverymen, and pilots will be hard-pressed to find work. And though those employees can retool, it's difficult to imagine how we might replace all of those jobs.

These more concentrated and automated modes of transportation also pose safety and security risks. Will driverless cars be able to respond as quickly and with as much agility as a human driver? Will pilotless planes be able to make split-second decisions that save passengers' lives in an emergency situation? And to the extent that these automated vehicles are connected to an operating system, could they be targets for nefarious hackers?

And all of this is to say nothing of the more trivial nuisances—nuisances like the electric scooters of San Francisco.

To experience the benefits of the transportation revolution, we'll need to address these concerns. We'll need to develop systems that provide the safety and security we need and expect, systems that allow us to trust machines to do the things we are so accustomed to doing ourselves. But will we be able to trust AI to drive us from here to there, even after the first few highly publicized autonomous car accidents take place? Will we share the road with autonomous 18-wheelers? Will we board pilotless planes or trust electronic engines to carry us a mile into the sky? Will we strap ourselves into *rockets* and allow ourselves to be *shot into space*, all to achieve a shorter commute to Shanghai? Will we be willing to suffer the mishaps along the way?

Time will tell if technological advancements in the travel industry will enhance our lives, if they'll provide us with an opportunity to ameliorate some of the transportation problems we've created over the years. We'll soon see whether modern technology will allow us to alleviate the unintended consequences of previous transportation disruptions—the wasted commute times, travel frustrations, congested city streets, and pollution. And as we move into this transportation revolution, we'll need to ask ourselves these questions:

1. How can we promote and utilize transportation technologies that alleviate the stresses of our daily commutes?
2. In what ways can modern transportation allow us to make travel more efficient, help us reclaim lost time, and alleviate the time pressures of travel?
3. How can modern transportation technologies integrate with modern communication and financial structures, so that we gain quicker, better access to the goods and services we want?
4. In what ways can modern technology provide opportunities for those without access to employment opportunities?
5. How can those in the transportation industry retool, so that automation doesn't detract from their livelihood?
6. How can we prioritize and safeguard modern transportation systems so they're less susceptible to disruption, whether by terrorism or otherwise?

We are going places, now more than ever before. Humans are taking over 1 billion international trips per year.[10] And as modes and methods of travel expand, it's worth asking whether each disruption is necessary and beneficial, or whether some of it is simply for the Birds.

10

COMMUNICATION

Perhaps no human endeavor has been affected by technology as much as communication. With the proliferation of social media, the rise of AI communication, and the expansion of holographic technologies, human interaction takes place in an increasingly brave new world. But does this new state of communication come with a cost?

It was April of 2017, and France was in the throes of a presidential election. The candidates were on the road, hustling down votes and trying to spread their messages just days before the country went to the ballot box. Time was running thin. The candidates' window to communicate their vision for France was closing. And far-left candidate Jean-Luc Mélenchon was working overtime to reach as many French voters as possible. As he did his best to seize every opportunity to stump, he double-booked himself. Actually, he septuple-booked himself. But this wasn't the result of an oversight—it was intentional. Mélenchon intended to be seven places at once.

The 65-year-old Moroccan-born politician who'd climbed the ranks of French government was enjoying a surge in popularity. He was also no stranger to technological means of communication. He'd used them throughout the campaign to advance his message. He was a regular You-Tuber, and his channel boasted hundreds of thousands of supporters. He enjoyed a robust Facebook and Twitter presence. And though he could have simply broadcast his stump speech to listeners at various venues (including their workplaces and homes) through his social media channels, as the election drew to a close, he hoped to make a bigger splash. He hoped to push the limits of modern communication and solidify his standing as the political darling of the younger voters.

On April 18, 2017, the leftist underdog took the stage in Dijon, while a hologram projection of him took the stage at six other venues at the same time. Through technology, he'd transcended the most basic human limitations of communication, the limits of time and space. He'd found a way to be in more than one place at once, and he'd given the listening crowd the illusion of embodiment, a novel and lifelike digital representation of himself. He'd used transHuman communication to advance his message.

It was a technologically savvy move, and it made a splash in the newspapers. There's no doubt his message reached more ears as a result, and it certainly gave him a boost in the voting. Ultimately, his bid for president was unsuccessful—but he'd successfully made a case that holographic stumping could be a viable campaign communication strategy. Politicians were paying attention because the political desire to project messages is rampant.[1,2,3]

Every politician seems to be connected to social media, including U.S. President Donald Trump, whose Twitter account is followed closely by supporters, political opponents, and late-night television comedians alike. (As of the writing of this chapter, on August 29, 2018, President Trump has tweeted 4,546 times since taking his oath of office.) And why do politicians project holograms, use digital avatars on social media, or appear on certain YouTube channels? Why do they use digital methods of communication? Simple. That's where the people are.

Modern mass communication platforms certainly have utility in many contexts (political, business, and personal), but does that mean these forms of communication are inherently good—or even necessary? Do they create new challenges, challenges we've not adequately addressed? And how can we take a human approach to balancing both the positive and negative aspects of transHuman communication?

The Proliferation and Effects of Modern Communication

Until relatively recently, the fundamental basis of human connection and communication has been person-to-person, one-to-one, face-to-face. Though humans have always searched for ways to transcend the limitations of in-person communication, to expand the reach of their messages, there was no true method of mass communication until the advent of the printing press in the 1400s. Another 500 years would pass before humans created an even more effective means of mass communication, the radio. Next followed the television, with its ability to beam lifelike representations of people across the world. Still, the power of mass communication belonged to the wealthy and elite—those with access to networks.

All that changed in the late '90s as digital messaging became possible via the World Wide Web. We were able to connect to and communicate with people around the world, even if they weren't tuning in at the precise moment we were communicating. We could send emails or put up webpages with relative ease and very little expense, spreading our messages at any time and to any place. It was a new first: Communication could transcend the barriers of time, space, economics, and—with the proper translation software—even language.

As technology advanced, our capacity to communicate followed suit. We gained access to more information than ever before and were able to make better-informed decisions. We could ask more questions, get more (and sometimes better) answers. We could share our thoughts and opinions with others more easily. But not all of it was as positive as we might have wished.

Chat rooms allowed us to share virtual spaces, to identify as whomever we wanted. A child predator could claim to be a 16-year-old girl, despite being a 45-year-old man. We pretend to be wealthy, famous—whatever our hearts desired. Then came social media, and it gave us digital faces, avatars. It expanded our ability to find anyone and everyone, to applaud or troll them. It gave us the ability to reach out and touch the rich and famous, the political movers and shakers. It allowed us to attack politicians or each other—or to spread fake news. And we could do it all with just a few clicks, from the comfort of our homes, in isolation.

By 2010, we'd developed more efficient, better ways of mass communication than ever before, and the world saw the effects. A revolution swept across the Middle East. Called the Arab Spring in the news outlets, some pundits deemed it a Facebook or Twitter revolution. And though some believe social media was used as a medium for organizing and mobilizing the uprisings, recent studies suggest that may not be true. Still, there can be no doubt that the use of social media platforms by online activists in Tunisia, Egypt, Libya, and Bahrain played a crucial role in the uprisings. For the first time, citizens in countries who limited the people's access to the press and other traditional media outlets had the upper hand. Through those social media outlets, the demonstrators found a platform to move beyond one-to-one communication and spread the news of the uprising to the outside world.[4]

And that was in the days when social media was young. Nearly a decade later, the use of the Internet and social media has grown exponentially. According to the Global Digital reports from We Are Social and Hootsuite, the number of Internet users in 2018 is 4.021 billion, comprising 55 percent of the world's 7.5 billion people. Nearly 3.2 billion people actively use social media (42 percent of the world's population). Almost 3 billion people (39 percent of the world's population) are mobile social users.[5]

Access to social media platforms and other forms of electronic communication has expanded our ability to share stories, communicate ideas, and connect with others. And in that way, transHuman communication (communication extending beyond the traditional person-to-person limitations) has had a positive effect on our lives. Those positive effects have come with their own set of challenges, though.

Depersonalization

As digital communication expands, we often find ourselves spending less time connecting with others in person and more times on our mobile devices. In 2014, researchers released a report, "The iPhone Effect: The Quality of In-Person Social Interactions in the Presence of Mobile Devices." The report reviewed the effect cell phones have on our individual interactions. It observed 100 couples engaged in 10-minute conversations and noted that when their phones were present, those individuals tended to fiddle with them, even while engaging in conversation. Conversely, when the phones were removed, the conversations between those same couples resulted in greater empathy. The report further noted that even when phones weren't "buzzing, beeping, ringing, or flashing" they were emblematic of the owner's wider social network and his access to vast amounts of information. As a result, even when with their significant other, the presence of their cell phone caused them to have "a constant urge to seek out information, check for communication, and direct their thoughts to other people and worlds." The article continued, "Their mere presence in a socio-physical milieu, therefore, has the potential to divide consciousness between the proximate and immediate setting and the physically distant and invisible networks and contexts."6

This connection between mobile communication technology and deteriorating personal communication is a thing readily understood by the younger generations. A published Elon University survey conducted by an undergrad student and advising professors asked students whether they believed technology negatively affected face-to-face communication: 92 percent said they believed it did, and only 1 percent believed it didn't (7 percent had no opinion). The survey also asked whether students noticed a degradation in the quality of conversation in the presence of technology: 89 percent believed there was, while only 5 percent disagreed (6 percent neither agreed nor disagreed).[7]

There can be no doubt that transHuman communication has its advantages, its uses, its utilities. But as the small sampling above—and, if we're honest, our own personal experiences—indicate, it can also be laced with potential relational pitfalls, pitfalls that might lead us to dehumanize one another.

A Brave New World: Human-Object Communications

As technological means of communication continue to expand, we find our communication isn't limited to interpersonal interactions. Increasingly, we're communicating with objects, machines, and the Internet of Things. We might ask Siri to give us directions to the nearest bar, or check with Watson on how our stock portfolio is performing. We might even ask Alexa to exchange a little friendly banter. And as our comfort level with Internet conversation increases, developers of modern technology hope to make those interactions even more lifelike.

Shantenu Agarwal, an IBM employee and resident Watson expert, took the stage at a recent conference to discuss "embodied cognition." It was developed as a way of giving customers the ability to interact in the "channel" with which they're most comfortable, he said, whether that be a person, an avatar, or a robotic representation. He then introduced Rachel, a near-lifelike talking avatar powered by IBM's Watson. Projected on the screen, Rachel's skin texture was indistinguishable from any other person who might appear on camera. Agarwal welcomed Rachel to the stage, and she indicated how happy she was to be in New York, stating that she'd never visited the city.

"Really, where are you from?" Agarwal asked.

"Have you ever heard of a place called the Internet?" she responded, lips curling into a smile. "Well, I've lived there all my life. I am a creation of Soul Machines. I can see you and hear you, and what makes me different is that I can respond to your emotions. I guess you could say I'm putting a human face on artificial intelligence."[8]

Rachel went on to explain that her current interest was credit cards and asked whether she could assist Agarwal in finding the right card, the card suited for his needs. In the moments that followed, they chatted away, having an almost human exchange. In fact, the exchange was so human, it was possible to forget that Rachel was not a human at all. She was simply the face of the Internet, speaking with Agarwal. In that communication, Agarwal shared his name, his needs, and even his credit score with the interface. What's more, he shared his image, the way his lips move when he talks and the way his brow rises and lowers—in other words, the micro-expressions underlying his communication. And all of that information was likely collected on a server somewhere, a server owned by IBM.[9]

A day is coming when we'll have access to our own Rachels, our own lifelike personal assistants. Those virtual assistants and advisors will likely be connected by broad neural networks, networks that allow for assistant-to-assistant communications. Together, those assistants will learn our preferences as we communicate with others, and they'll mirror those preferences. They'll learn to answer questions better and more efficiently—and to anticipate our needs, especially as computing power grows.

transHuman communication with the Internet (whether by way of digital assistants or otherwise) has no foreseeable limitation. In fact, with more development, more refinement, it's not difficult to see a world in which digital advisors will supplant many of our day-to-day jobs. They may replace teachers in schools or universities and have direct interactions with our children and grandchildren. They may become our conference speakers, financial advisors, and insurance salespeople. They may functionally supplant any knowledge-based, customer-driven interaction. As they do, they'll do it in ways that make it virtually impossible to recognize that we're interacting and communicating with machines. And yet, these interactions will continue to undermine human-to-human interactions.

These same methods of communicating may transform human communications, too. Our human modes of communication won't be limited by time or space. We'll no longer be confined to in-person conversation or to exchanging words on a screen. We won't need to use cartoonish emojis to try to bring nuance to a text, email, or tweet. Instead, our own digital avatars will transmit our messages, and those digital avatars will be just as lifelike as Rachel. They'll use micro-expressions to bring our thoughts to life. What's more, they'll be able to transcend cultural and language barriers. Powered by the Internet and robust algorithms, our messages and expressions will be translated on the fly, perhaps to multiple people at the same time, with all of the appropriate cultural nuance. Through these modes, we'll be able to simultaneously speak all the languages of the world yet maintain our national customs and singularities. The applications of these forms of technology might just revolutionize commercial and political discourse.

To complicate matters, humans won't just be communicating avatar to avatar, or through human-like Internet interfaces. Increasingly, we'll be communicating with objects: "Hey, refrigerator, ask the grocery store to deliver milk." What's more, our objects will communicate with other objects without any need for human intervention. When it senses we're running low on milk, our refrigerator might order it from the refrigerated milk case at the local grocery store. That grocery store milk case might communicate with a drone to deliver it. That drone may communicate with your door lock or garage door, allowing it to enter your home and deliver your groceries.

Object-to-object communication won't be confined to consumer interactions, either. Smart cities are already springing up where city traffic sensors, lights, energy management systems, security systems, and other city services speak with each other. Imagine a world where a traffic sign at a busy four-way stop communicates with the traffic light three miles down the street, telling it that a heavy volume is coming and that the light should let more cars through the intersection to accommodate the increase. Imagine a reservoir sensor communicating with a treatment facility if it discovers the water is contaminated.

In these ways, technological advances in communication offer so much that could make the world a better place. But with increased digital communication comes increased risk.

The Risks of transHuman Communication

An increasing dependence on social media and digital devices has negatively impacted human connection and may be linked to the decreased empathy in today's world.[10] And it doesn't take much to imagine that with decreased human-to-human connection and decreased empathy comes increased division and vitriol. (Have you looked at social media lately?) With increased division and vitriol comes an increased risk to the security of our communications.

Mass electronic (and holographic) communication opens all of us—individuals, markets, and governments alike—to security challenges. Internet providers, social media networks, digital assistants, and every other communication-hosting platform collects a wealth of personal information from users, and some obfuscate their true motives by putting on a friendly face. That data allows those companies (using advanced AI algorithms and other methods of communication) to target users with specific ads and marketing campaigns. What's more, with increasing regularity, these same companies can gather and analyze the emotional and ideological data of each person connected to this vast network of information. And if a cyberattack led to another data breach, a whole cache of personal information—including facial recognition data, emotional data, and even speech data—could be compromised.

These methods of transHuman communication subject platform users to fraud and manipulation, too. On occasion, those risks are known and quantifiable. Consider the Ohio State University study, which indicates "fake news" and mass voter manipulation via social media may have contributed to Donald Trump winning the presidency of the United States.[11] And as technology advances, it's not difficult to imagine an age when that fake news might be even more compelling. For instance, in an age where holographic campaigning is becoming more commonplace, a place where political candidates might appear at several places at the same time, would it be difficult to create a false hologram of a political opponent? Would it be all that difficult to create actual fake news using your opponent's likeness?

Consider, too, the economic risks of these new methods and modes of communication. As digital assistants increase in intelligence, more jobs traditionally reliant on human-to-human communication will go the way of the dodo bird. The trend will create a boon for companies that design and service these digital assistants, while creating a drag on many sectors of the human economy. Entire industries may be eliminated, while those with better access to technology may increase their wealth disproportionately. The shift of capital could lead to higher corporate concentrations of wealth, with the impact being felt in greater proportion by the everyday worker.

Technology, then, opens transHuman communication up to significant risks—risks for which we're not yet prepared. But for all the risks posed by these technological advances, some can be mitigated by these same sorts of technology. In this brave new world, all participants in the communication network form a single global body. And just as the body has white blood cells to fight viruses and attacks, resources need to be poured into protecting our communication networks. We'll need automated systems that detect security threats, data breaches, and fake news, that communicate with the other cells of the network so it can organize and defend itself—and learn from attacks.

But in this new environment, cybersecurity systems to protect our modes of communication won't be enough. We'll need to ensure, too, that all of our human-to-human communication doesn't happen via electronic means exclusively—means that don't place a premium on empathy, understanding, and civility.

Prioritizing Human Connection in Modern Communication

When we examine the changing landscape of mass communication, the issues are clear. transHuman communication has the potential to erode the quality that forms the basis of all our human-to-human engagements: connection. So, when examining modern communication methods, we must ask ourselves how we can balance our need for mass communication with our need for peer-to-peer connection. Where can we start? Perhaps by asking these kinds of questions:

1. In what ways can I prioritize person-to-person communication over methods of electronic communication across digital networks?
2. In what ways does digital communication help me get tasks done, reconnect with old friends, or get information I otherwise wouldn't have? In what ways does it invite me to attack others, show a lack of empathy, subject me to data breaches, and open me up to manipulation by targeted advertisements?
3. How can we reduce (or even eliminate) cybersecurity threats to our social networks and stop the proliferation of manipulations like fake news? Are there technologies designed to better protect those networks, technologies in which I can invest?

In the transHuman environment, humans will continue to communicate in virtual spaces, and this isn't all bad. Communication is the most natural act of humans, and when engaged properly, it can lead to a beautiful exchange of ideas and stories. It can lead to greater understanding.

But, until systems are in place to minimize the risks—psychological, economic, and cybersecurity risks—we should proceed with caution. And the next time we see a politician stumping via hologram, or the first time the Internet attempts to communicate with us via a lifelike avatar, we should ask some probing questions about the nature of what it means to remain wholly human while still communicating in the most advantageous transHuman ways.

11

COMMUNITY

Safety and security. Overcrowding and congestion. A lack of infrastructure and services. These are the concerns of modern community developers. And as in so many other areas of life, many have turned to technology in an attempt to solve these problems. But is the application of technology the best answer to these concerns?

In the middle of the 20th century, President Juscelino Kubitschek of Brazil held a city-planning design competition. He hoped to find the design for a new capital city, one worthy of housing the Brazilian government. He wanted a beautiful city, an organized community without the urban sprawl and blight of Rio de Janeiro. And though it'd be an audacious task to build this kind of city from the ground up, Kubitschek was committed.

The competition began, and among those who answered the call were city planner Lucio Costa and his collaborator, Oscar Niemeyer. Costa submitted his plan, which included modernist designs and the application of the newest technologies. There were high-density buildings for both residential and governmental use. The lines of those buildings were clean and modern, and they made use of natural light. The structures rose above wide boulevards to accommodate the flow of automotive traffic. Those boulevards were laid out with modernist precision, too, designed to maximize traffic flow without the need for stoplights or traffic signals. There was ample green space, which gave the city a more spacious feel. Everything was avant-garde; it was the city of the future.

Costa's plan stood out as the clear winner and was chosen as the design for the new capital, Brasilia. Construction began in an open field, and in less than four years, what had once been nothing more than a set of plans became a full-fledged city. And for a while, it was a dream come true. The government moved its operations to Brasilia and people descended on the city. But as they did, the problems with its design became apparent. Brasilia was designed as a concept, a city for the sake of a city. And though it was organized well for an efficiently functioning government, it wasn't designed with human habitation in mind.

For the average working-class citizen, it was nearly impossible to live in the new capital city. Housing in the ultra-clean, ultra-efficient, ultra-modernist utopian community was expensive. To make matters worse, areas of commerce—marketplaces and retail centers—were limited. Entertainment was even more limited. And as a result, the city's already cold vibe grew colder after working hours. So, more dense urban areas sprang up on the outskirts of Brasilia, areas where the activities of everyday life occurred. But those areas weren't as well planned. In fact, they became every bit as dingy and blighted as those of Rio. But against the backdrop of the pristine and minimalist government city, that blight was, perhaps, even starker.[1,2]

In a 2010 *Reuters* article, it was reported that Brasilia suffered from violence, lack of infrastructure, and terrible sanitation issues. And plans to address the infrastructure and sanitation issues were put on hold in the midst of corruption scandals in the government. In that same article, it was reported that "Niemeyer, a famous architect and communist, said he is disappointed at the huge wealth gaps that now scar the capital."[3]

Brasilia was designed to advance a utopian ideal, and it used the most advanced technology and design elements of the day. But the story of Brasilia is a cautionary tale. Whether designed from the ground up or in a piecemeal fashion over years, city planning that fails to take human needs and activities into account will ultimately end in failure. It may end in overcrowding, excess pollution, traffic congestion, or urban blight. It may end in increased crime, violence, security risks, and wealth disparity.

All of this raises the question of how we can use technology (and design) to address human needs in city planning, instead of employing them in ways that create negative, unintended consequences.

The Individual as the Fulcrum to the Community

As we've entered the Fourth Industrial Revolution—the technology revolution—populations continue to expand. With that expansion, the world has entered a cycle of urbanization. As a result, we have greater concentrations of people living in smaller areas than ever before. Some of those cities have grown to the extent that their annual GDP is larger than that of a small country. In fact, in some instances, our urban centers have grown so large and influential that the mayors have more political sway than governors. Why? Because the higher concentrations of people have made power more localized. And though this trend is neither good nor bad *per se*, for the most part, it has occurred without consideration of the effects on the humans that populate those cities. It's a troubling thought.

In the midst of this Fourth Industrial Revolution, we cannot continue the trend of urbanization without keeping an eye on the community's human inhabitants. If humans represent the highest and best technology, then everything we design, implement, or invent—the algorithms, artificial intelligence, networks, robotics, and systems of communication—should be used to improve *our* lives. And this is true whether we're discussing the application of technology to water, food, education, or community planning. In other words, community planning and design should first consider the people who live in those communities.

When it comes to community planning, putting the human first, making him or her the fulcrum of the community, may require us to apply *less* technology, not more. We may need to organize smaller communities, communities that allow for smarter living experiences, instead of creating cities that prioritize greater concentrations of people living together. After all, according to Adam Okulicz-Kozaryn's study entitled "Unhappy Metropolis," residents of large cities are generally less happy than their smaller-city counterparts.[4] And if this is the case, we may not need to implement smart technologies for the sake of making big-city life easier. We may not need to create better, more advanced traffic systems, utility distribution hubs, or waste disposal methods to deal with dense urban populations. Instead, we may need to use technology to decentralize the city altogether, to create smaller but more connected communities.

In countries like Switzerland, small and efficient cities dot the countryside. (Is it any coincidence that Switzerland is consistently ranked among the happiest countries?) Many of those cities are designed to be traversed on foot or by bike. Many have mixed-use areas of commerce and housing, which reduce the need for vehicles and cut down on the resulting pollution. Long, isolated commutes to work and home are unknown. It's an approach that elevates the human, that takes a practical approach (instead of a technological one) to curbing congestion and pollution and the problems that come with urbanization. And if these benefits weren't enough, this approach helps to create communities where human interaction and connection are not only encouraged but also required. In other words, these smaller cities serve to meet a core need of the human— human engagements—and for the most part, do it without the need for too much technological innovation.

This is not to say that technology can't be employed to make cities better. After all, everyone won't opt to live in the less urbanized areas in the Swiss countryside, and even in those communities, technology can help connect the residents with each another. But to the extent technology and design are employed in community development, those means must be used only to the extent that they enhance the lives of humans populating those areas.

What might that look like?

There can be no doubt that the cities of the future will be run by high-powered operating systems, and those systems will be connected to secure clouds. The residents of that city will be able to interact in real time with community service providers. As they go about their daily activities, they'll schedule transportation needs, tend to educational needs, pay their taxes, interact with their fellow citizens, and maybe even vote, all through the city operating system.

These operating systems will also handle complex logistical needs of the community, all without requiring input from the residents. Objects with smart sensors—traffic lights, emergency response services, even disaster relief services—will communicate with each other, reacting to the needs of citizens instantaneously. They'll minimize congestion, automatically rerouting traffic when needed. And because citizens will be connected to the city through their own devices, their lives will be enhanced, especially from a safety-and-security perspective. Local police departments will be able to identify those involved in accidents, those who've been the victims of crimes, and those who've perpetrated those crimes, all as a result of discrete (and secure) digital identities. Within seconds of an accident, first responders will be notified through smart traffic sensors. And all of this will be possible because of the network of connected municipal objects, all working together.

Human Interactions in the City, with the City

Unfortunately, one of the by-products of the Third Industrial Revolution—the revolution that gave birth to urban sprawl and the rise of the disconnected suburb—is the fact that humans have become more and more isolated.[5] Ironically, in this connected age we live in, we're more physically disconnected from each other and from local places of commerce than in the past. The reason is no surprise. The rise of technological communication tools, the digital marketplace, and the expansion of outlying suburbs make it possible to share ideas, engage in commerce, and even attend community events without sharing physical proximity with another person.

As we mentioned in our discussion of communication, the need for human connection is one of our most important needs. What if communities used technology in ways that restore human connection at the local level? What if they used technology to proactively hedge against isolation?

Imagine a future in which greater connectivity to city operating systems allowed the city to detect the needs of its citizens and connect them to other citizens who might meet those needs. Imagine a matchmaking model, but not one that initiates romantic connections. Rather, imagine a system that operated at the municipal level, one that connected people based on either affinity or need. And this community system could introduce people to people, people to businesses, or businesses to businesses, all by way of an opt-in service using an affinity- or need-based algorithm. This kind of system could improve the citizens' social networks, allowing them to find jobs, houses of worship, or local bars *where everyone would know their names*. In this way, smart communities could help promote meaningful relationships among its members.

These connections won't just be limited to people in the local community, though. Communities of the future will also be connected to each other. Cities with certain specialties, products, and services will initiate commerce with each other. They'll share information about crop availability or unique tourism opportunities. They'll inform neighboring cities and their citizens of employment opportunities, opportunities that might be better suited for their neighbors. They'll share information about local clubs, athletic events, or organizations with members of outlying communities, in hopes of bringing larger groups of people together around common ideas, passions, or needs. In this way, the communities of the future can forge even better human connections.

It goes without saying, though, that these kinds of connected communities will create unique challenges. Connected cities will gather immense amounts of data on their members—where they travel, the stores they visit, the number of accidents in which they've been involved, whether they've been the victim of a crime, whether they've perpetrated a crime, the people with whom they interact, and so on. This data will give municipal leadership great insight. Still, municipalities will need to go to great lengths to protect that data from theft or misuse, and corrupt politicians may seek to use it to manipulate or influence the citizenry. What's more, corporate entities within the municipality may be willing to go to extraordinary lengths to harvest that data, data that would give them insight into consumer choices. But if systems can be employed to ensure against the misappropriation, misuse, or manipulation of that data, these kinds of operating systems might greatly enhance the lives of the citizens.

The Buildings of the Future

Technology won't simply be used to connect humans to each other or to their cities, though. In this Fourth Industrial Revolution, technology may be used to lower the costs of living in those cities, a primary transHuman code objective. With the proliferation of 3D printing, printing construction materials for use in housing has become a reality. These materials are cheaper and often more durable than traditional building materials. They'll allow us to cut back on our use of both renewable and nonrenewable resources. The construction of the future will rely less on timber and petroleum products, a net positive for the environment.

In some places around the world, those printable materials are already being used to create affordable and modern housing. Icon, a "construction technologies company dedicated to revolutionizing home building," is actively asking, "What if you could download and print a house in 24 hours for half the cost?" And this isn't simply a hypothetical question. In March of 2018, they printed their first prototype home in Austin, Texas. And though the technology advanced by Icon is cutting edge, they're employing it in truly unique and human ways; they're attempting to use the technology to create communities for the most impoverished people in the world. According to the promotional video found on their website, "The first community of homes will be printed in El Salvador, with additional locations soon to follow."[6]

3D printing technology has also been used to create furniture—furniture that's cheaper, more durable, and ultimately recyclable. Using printable materials to drive down the costs of homes and furniture could lead to significant consumer gains while simultaneously reducing waste within the community.

The homes of the future won't just be printed from renewable and recyclable resources, though. They'll be more efficient and reconfigurable, allowing us to use less space while living more comfortably. Transform-able furniture systems will seamlessly adapt any space to the appropriate activity, so that one room can have multiple uses. Imagine a bed that auto-matically folds into the wall; a dining room table that transforms into a desk; a wall that moves into the adjacent room—all to provide more space for entertaining, or creating a home office, or whatever. In this way, space might become more functional and usable while also being cost-effective for the average, or even low-income, consumer.

The application of modern technology to architecture will have an impact on our city skylines, too. By using more printable materials, we'll reduce the carbon footprint of our average skyline as we produce less steel, concrete, wood, and other consumable materials. Through the use of LED lighting in our modern buildings, we'll reduce our need for electricity by as much as 80 percent, decreasing our consumption of fos-sil fuels and other alternative energy sources. And by using some of the same modular furniture and adaptable spaces, a business may transform an office into a community gathering hall with the touch of a button, thus creating more places for the people of a community to connect.

These may seem like ideas whose time has come, but as with any applica-tion of technology, there are resulting risks to human flourishing from the application of this technology. Without intentionality—intentionality like that used by Icon—many of these applications may disproportionately help the wealthiest among us. They may also continue to encourage the growth of already dense urban areas, leading to further congestion and decreased happiness. What's more, modern construction technologies may eliminate thousands of man-hours per house, adversely affecting many blue-collar laborers. So, though these technologies may alleviate many of the current issues relating to community development, before we apply them, it stands to reason that we ask these most important questions: Do the benefits derived from these technologies outweigh the negative impacts? Do these technologies do the most good for the most people?

The transHuman Code and Communities

There can be no doubt: New challenges come with the global increase in urbanization. Increased pollution, safety and security concerns, the lack of human connection—these are the challenges facing us we push into the future. And though some of these issues may be solved by way of technology, the application of less technology may be the first step: building smaller communities; living within closer proximity to our places of work, our local bars, our houses of worship; cultivating community with local artisans, farmers, and friends.

But as we apply technology to our communities—as we invariably will—let's keep our eyes on the unintended consequences. Let's seek to answer these questions:

1. How can we design cities that are both efficient and livable?
2. In what ways can we enhance citizen safety by using secure, automated systems?
3. How can we connect citizens to those automated systems while still protecting their privacy and personal data?
4. How can technology facilitate citizen interactions and connections?
5. In what ways can technology be used to make housing more affordable, while still providing jobs for less-skilled laborers?

6. Can we use purposeful design to create functional yet desirable multiuse spaces?

These questions cannot be answered by technologists, architects, and city planners alone. They require *our* input, the input of the people living in the cities. Without that input, we'll apply technologies in less-meaningful ways, ways that create cities more like Brasilia and less like the communities of Switzerland. But if we take the transHuman code seriously, if we apply technology only to the extent that it prioritizes and maximizes human needs, we can create community systems that encourage human connection, facilitate local transactions, and maximize the security and safety of the citizens. We can create affordable, efficient, safer, more secure communities.

Prioritizing human interactions, connections, and access to human needs—these should be the goals of community planning and development. And as we employ technologies to meet those goals, let's be deliberate. Let's ensure we're serving the best interests of the community. Let's ensure we're keeping our eye on the individuals in those communities: us.

12

EDUCATION

In 1987, the German technology company Fraunhofer-Gesellschaft launched an in-house research project on digital music compression technology, code-named Eureka. Two years later, the company received a German patent for their discoveries. But it would be seven more years of trial and error before Fraunhofer's research produced a viable, proprietary representation of all its work. They called it MPEG-1 Audio Layer III—mp3 for short.[1,2]

From 1994—the year in which Fraunhofer released the first mp3 decoder, making it possible for people to listen to digital music on their computers—to 1997—the year in which www.mp3.com began offering thousands of free digital music files via the Internet—the landscape of the music industry was being turned over like a field in early spring. Then Napster came along in 1999, and the industry transformation scaled to global proportions. Suddenly, music consumers anywhere had choices—and they quickly made their preferences known. No longer did they have to purchase an entire CD to listen to the two songs they liked. And no longer did they need to own multiple CDs to create a unique, personal mix. In a matter of minutes, they could listen to specific songs and self-made mixes from the comfort of their homes or workplaces. Best of all, it was free.

We know what came next. Record companies cried foul, claiming that free downloading and sharing of digital music files was not only copyright infringement—it was also putting a significant dent in album sales. This much was true. But instead of seeing an opportunity, their legal departments sought to make examples of the biggest perpetrators. At the top of the list was Napster, which was eventually shut down. By that time, however, a new world had already emerged. Consumers had tasted its fruits and their demands had shifted. The music industry fought to reign in consumers' demands—and even return them to a consumption model that had diminishing appeal.

The scuffle continued until, eventually, the music industry had a new eureka moment and realized its best course was to adapt to the overwhelming demand, rather than fight to control it. They had help from one particular pioneer named Steve Jobs, who introduced the world to iTunes—and its counterpart, the iPod—in 2001. A little more than a decade later, digital music sales had equaled physical album sales; today, digital sales have surpassed physical album sales.[3] Since then, the music industry has worked hard to find the right balance of offerings that consumers desire. Most importantly, while consumers listen to music, the industry is now intently listening to them.

This story serves as a contemporary parable for the future of education.

Where human learning is concerned, we are living in historical times. There is more venture capital invested in education technology (or edtech) than ever before, with the global total surpassing $1 billion in 2012 and reaching $9.52 billion in 2017, "across every market involved," according to both *Forbes* and Metaari, a Seattle-based area market research firm.

Private funds are also streaming in, with all 10 of *Forbes*'s richest people directing portions of their wealth toward educational developments and endeavors in the last decade.[4]

What this amounts to, at the very least, is that we now have—at our fingertips—access to the most efficient, customizable, quantitative learning tools available. These tools are changing the methods and means by which we grow as individuals and progress as teams. Many such tools are already widely available, like iTunes U, which allows anyone to listen to an expert explain anything—from ancient conflicts to the most important works in literature—and the math- and science-focused Khan Academy, whose mantra is to offer "a free world-class education for anyone, anywhere." We've made major strides in the past five years alone. And yet, we've only scratched the surface of what education can look like and, most importantly, what it can provide—not only for the typical swath of kindergarten to college-aged students but also for learners of every age, and every reason, around the world. The promises of better environments, superior customization, and greater overall effectiveness are on the near horizon and, in some ways, already here.

Unfortunately, as with any advancement in any long-established industry, there is always pushback. The old guards of traditional education—and its advocates—are slow to adapt, if not altogether resistant. The static conventions they uphold, such as accreditation standards, tenure tracks, and standardized credentials, are keeping some of the best tools on the fringes of the industry, relegated to largely tertiary environments. As a result, much of what is available right now is not yet offered to the vast majority of students. Or, if it is, it's not offered by traditional educational institutions. The tide is shifting, but we're left to wonder if it's shifting fast enough.

Those who see the new learning opportunities as a threat are like those in the music industry who saw the advent of the mp3 in the same light. Just as musicians (and those who paid them) envisioned their livelihoods fading away with the rise of personalized digital music, some committed educators see their livelihoods fading away with the rise of personalized digital learning.

Though it's tricky to defend this position today, some still refuse to leap from the sinking ship called traditional education, which, according to Anant Agarwal, CEO of the nonprofit MOOC platform edX, "hasn't changed in hundreds of years."[5] Such educators are not much different than the litigious record company execs of the late 1990s, who, instead of moving nimbly to adapt to the industry's (and humanity's) changing standards, sought to hunker down and protect an outdated paradigm.

In this sense, the story of the mp3 also serves as a timeless lesson.

Progress to any degree, whether personal or global, has always come on the heels of a battle between convention and innovation; it's that timeless tug-of-war between what must remain and what is open to change. Convention should be valued for the foundation of growth it provides: stability, predictability, and an environment for establishing best practices. Convention works—until something better comes along. Then, in what seems like the blink of an eye, convention becomes the biggest hurdle to implementing an innovation—and, consequently, to progress. Both convention and innovation are necessary, but they must take turns being subordinate. Right now, in the education industry, innovation must reign.

We are now seeing that the "doomed" fate some early objectors so confidently predicted for online education has not, and will not, come to pass. Quite the opposite is occurring: Consumers of education are gaining greater freedom, as well as stronger footholds on the conditions and results they seek. While some people still hold to a belief that anything beyond traditional, classroom-based education will remain secondary, many today have changed their minds. They've seen the new future unfolding before them at a rate few predicted—including the proponents of innovation—just a few years ago.

Today, most educators in the U.S.—91 percent, according to the 2016 "Teachers' Dream Classroom Survey"—agree that technology helps facilitate better learning, but only 16 percent of these same teachers give their schools an A for incorporating technology into the classroom.[6] The number is poised to climb as more pioneering educators, like MIT's late president, Chuck Vest, step forward and embrace the future through initiatives like the groundbreaking OpenCourseWare project, which offers free materials online (syllabuses, tests, etc.) for thousands of MIT courses. "The project was," explained *The New York Times*'s Ashley Southall, "a model for other universities in developing so-called massive open online courses" (today, simply known as MOOCS).[7]

In a 2011 interview, Richard DeMillo, director of the Center for 21st Century Universities at the Georgia Institute of Technology, summed up the impact that pioneering advances like OpenCourseWare had on the future of learning: "Putting every course online, free, showed that the value of a degree from MIT was not contained in the lectures and the exams and homework. It's contained in the experience of passing through the network of MIT scholars." What was his conclusion (and that of many progressive thinkers in education)? "[W]hy hang on to what should be shared widely?"[8] Furthermore, why confine learning to a predefined time, place, and pace when a world-class education can be offered, pursued, and evaluated whenever, wherever, and however learners desire.

These questions frame the new spirit of learning and cultivate today's education landscape. The hurdles of convention are nothing new—and some are high hurdles—but they must be dealt with nonetheless. "For 150 years, formal education has adopted an 'inside-out' mindset," explains David Price in *Open: How We'll Work, Live and Learn in the Future.* "Schools and colleges have usually been organized around the needs of the educators, not the learners . . . The new landscape presents a significant upheaval. Inventors and researchers are increasingly working independently outside academia . . . Learners also find themselves in the driving seat because formal education is no longer the only game in town for those eager to learn." How educational institutions adapt, says Price, "will largely determine their survival."[9]

This vital adaptation is great news for those of us who love to learn, need to learn, or lead learners for a living—which, we must admit, is all of us. The education landscape unfolding before our eyes will not be without its growing pains. But what we can expect is an environment in which high-quality learning is de-privatized—is made collaborative, accountable, and relevant to the desires and needs of people the world over. Best of all, this sort of model of education puts the keys to our learning potential in the hands of each individual. For the positive progress to continue, we need only to use the keys wisely and in a concerted fashion.

Perhaps we'd be wise to begin by recalling the quote widely attributed (though not confirmed) to Buckminster Fuller: "You never change things by fighting existing reality. To change something, build a new model that makes the existing model obsolete." If our aim is for education to benefit all the citizens of the globe, it seems we'd be better suited using our partnership with technology to "build a new model that makes the existing model obsolete."

We're getting there—*tinkering* our way there is the better description—but we can do much more, on a much larger scale. We cannot forget that we are both the consumers of education and its sole benefactors. Imagine transitioning current education to an environment where open data sources like YouTube and Wikipedia are just the tip of the iceberg for ongoing, accessible learning. Also imagine, in that same environment, a new way of qualifying and quantifying a person's education. For instance, what would happen if education became wholly molecularized?

Without the need to sit in a classroom, virtual learning enables education to not only be broadened to global platforms but also become focused down to the individual's learning needs from moment to moment and year to year. In this case, we would no longer need to learn in multiyear chunks, going from elementary school to high school, then to college and graduate school, spending decades earning standardized certificates—which, to date, are nearly always required in a job search. Instead, we would each learn according to the human needs and desires we have in each stage of life, with virtual tutors available 24/7 to aid human teachers, and an AI algorithm that paces with our education trajectory, testing whether we have learned, tracking what we have learned, and then quantifying the value of our education via a global ledger. We would, theoretically, no longer have to surmise whether a person was qualified for a job, role, or elected position; our technological partners would make this very clear. This would not necessitate the removal of human discernment in hiring, choosing, or voting. It would, however, remove unqualified people from the mix, placing the focus on the intangible qualities of each individual, such as empathy, ambition, and creativity.

Such a personalized, trackable approach is already viable in career settings that require a constant "update" to human knowledge, as laws or technological capabilities change. Many in medical and legal professions today are required to undergo regular (usually annual) education upgrades, as fundamental changes to law and medicine are ongoing. What if this approach was expanded, not just to the formal education of children but also the continual learning that occurs throughout adulthood; not merely for career improvement (or in many cases, to fulfill protocols) but also for acquiring new skills needed, or wanted, at various times and seasons.

"The education system is full of opportunity," says investor and entrepreneur Vinod Khosla. He continues:

> But as of yet, there is no simplified, decentralized path for every citizen of the globe to access the world's knowledge and be credited (or accredited) for what she learns from the day she is born until the day she dies. Perhaps this is the new blueprint we ought to be composing and building.[10]

"We should do away with the absolutely specious notion that everybody has to earn a living," wrote Fuller. "It is a fact today that one in ten thousand of us can make a technological breakthrough capable of supporting all the rest. The youth of today are absolutely right in recognizing this nonsense of earning a living. We keep inventing jobs because of this false idea that everybody has to be employed at some kind of drudgery because, according to Malthusian Darwinian theory he must justify his right to exist. So we have inspectors of inspectors and people making instruments for inspectors to inspect inspectors. The true business of people should be to go back to school and think about whatever it was they were thinking about before somebody came along and told them they had to earn a living."[11]

As it applies to the marriage of humanity and technology, the real questions around the future of education have more to do with the acceptance and adoption of innovation within the institutions that still largely govern how we learn from (approximately) 5 to 22 years old. Humanity at large, today, seems plenty eager to explore and learn to every degree imaginable—just not *for* every degree imaginable. It's time we, the planet, begin addressing the bigger questions:

1. How should educational institutions re-engineer the use of education to better suit the world we live in?
2. How can technology be used to continue opening paths of education that rival, or even surpass, the highest institutional forms available today?
3. How can students of every age use their consumer power to steer education to where it best suits the needs of the world?
4. In what ways can educators become a voice for humanity, rather than merely a voice for their institutions?
5. Can the formal education system that exists in many developed countries be changed—not in decades, as some suggest it will take, but in the matter of a few years? If so, how?

As we seek to address these and other important questions, let's remain open to learning from any corner of the planet and any person on it. To aggregate all global knowledge—not just in books, manuals, and libraries but also in our minds, hearts, and experiences—would be something humanity has never achieved before. Today, it is possible for us to create this resource. We just need to be open to finding the path and working together to pave it and protect it. There would be no more powerful tool on earth than open access to the collective mind of humanity.

13

GOVERNMENT

We've examined how we might ensure food and water security for every individual on the planet. We've explored how expanding technologies in the fields of medicine, energy, and transportation might bring about greater levels of prosperity and happiness. We've looked at the effects of technology on economics, including the job and finance sectors. And we've asked how technology might be used to enhance the security of our citizens. Historically, the government has played some role in each of these sectors. So, to an increasing degree, our current government structures will need to grapple with technological disruptions to these sectors and will need to sort through regulatory structures to ensure the interests of the human will be met.

That said, government structures will not simply deal with the technological disruption of industries as a tangent. Disruption will enter into our democratic process and make our political systems more granular, more decentralized, and more effective for the people. As it does, we'll need to ask how we can protect democracy itself from sophisticated technological attacks.

In March of 2018, whistleblower Christopher Wylie sat with British News outlet *The Guardian*. The topic du jour was Cambridge Analytica, the political consulting firm cofounded in 2014 by Alexander Nix, *Breitbart* editor Stephen K. Bannon (who became an advisor to President Trump), and hedge-fund billionaire and Trump supporter Robert Mercer. The company had been thrust into the spotlight after its questionable practices came to light, and Wylie was willing to tell all.

In the interview, Wylie, the former director of research, shared Cambridge Analytica's unique plan for obtaining the private data of millions of Facebook users. After forming the company, the firm retained Aleksandr Kogan, a senior research assistant at the University of Cambridge. According to Wylie, "Kogan offered us . . . something that was way cheaper, way faster, and of a quality nothing matched."[1,2]

What did he offer? An incomparable data scraping method.

Kogan designed an app called thisisyourdigitallife, which was rolled out as a personality quiz on Facebook—and which harvested users' data. But it didn't stop there. Through its advanced algorithms, it reached beyond the users who voluntarily took the quiz, harvesting the publicly available information of their Facebook friends and acquaintances. Using these methods, Cambridge Analytica was able to gain access to upward of 87 million Americans in the months leading up to the election. Using that scraped data, it created detailed user profiles. The firm then targeted select profiles with messages and advertisements, in an attempt to influence the election toward Republican outcomes.[3,4]

What kinds of messages and advertisements?

In an interview with BBC, Wylie indicated that Cambridge Analytica targeted users with fake news from sources that appeared to be credible news outlets, though not within the mainstream media. In this way, the firm sowed the seeds of distrust in the mainstream media, as targeted users began to ask why those stories (the false and misleading stories spread by Cambridge Analytica) were not appearing on the likes of CNN, the BBC, and other more traditional news sites. It was a "full-service propaganda machine," Wylie said, a way to influence the election.[5,6,7,8]

Cambridge Analytica had used psychographic data and manipulation to influence voters on social media, and to some degree, they'd been successful. It was a novel, brave new political disinformation campaign, a method meant to undermine the free press—and no one could have seen this sort of voter manipulation coming. Right?

Wrong.

Michal Kosinski, a researcher and assistant professor of organizational behavior at the Stanford Graduate School of Business, indicated in an interview that "[he's] been warning about these risks for years." He continued by saying, "Our latest research confirms that this kind of psychological targeting is not only possible but effective as a tool of digital mass persuasion."[9]

Through complex and high-tech means, Cambridge Analytica had skirted traditional media outlets and influenced politics through back doors. It was a highly targeted, highly technical process, made possible as a result of our decentralized media outlets. But this method of political influence will only increase as we push into the Fourth Industrial Revolution. As government systems themselves become more decentralized, as individuals begin voting on a cadre of discrete issues, partisan hackers and politicians will have more and more opportunities to influence and manipulate the vote. They'll have greater incentives to undermine and manipulate free speech.

What does a decentralized government look like? How could the application of technology change democracy? And how can we protect it from disinformation and manipulation of the powerful, wealthy, and elite? These are the questions we will need to answer as technological disruption comes to the governmental sector.

Government in the transHuman Era

Historically, governments have taken a top-down, highly centralized and paternal approach to governing. Aligned with ancient Greek principles, democracies have concentrated power within large governmental bodies and employed representative professionals—politicians and their staff— to run those governments. And in the days when the citizenry was spread across large swaths of land, in the days when they had little access to sophisticated information, this made sense. We sent representatives to the capital and trusted them to analyze information and voice our concerns. And even if they were sometimes persuaded by lobbyists and special interest groups, we still trusted them to vote as we would: with our best interests in mind.

For the most part, these structures are still in place. We're governed by documents written hundreds of years ago by men responding to the technological limitations of their own day. Technology has dramatically reshaped our access to education and information; it has dramatically reshaped the ways we communicate. As a result, many of us have developed our own areas of expertise—at levels that might, at times, outstrip those of our governmental representatives. We may know more about health care, business regulation, taxes, or cybersecurity than our representatives in faraway capital cities. And as a result of modern forms of communication, we can voice more of our opinions and make more of our own decisions. This raises the question—do we still need the bloated bureaucracies of yesteryear, those slow-moving and cumbersome democracies?

As technology advances exponentially, we're seeing the earliest signs of its disruption of traditional large government systems. Whereas the 20th century was marked by more centralized governments inefficiently providing overpriced suites of services, the 21st century will be characterized by direct access, direct voting, and a more efficient delivery of services.

We've already begun to see more and more connected government interactions and services. Through online portals, citizens are paying their taxes and accessing government-provided utilities. They're registering to vote (and, in some countries, actually voting), changing their names, registering vehicles, and filing complaints. The streamlining of these services has simplified our lives and provided budgetary savings for our governments. The Fourth Industrial Revolution will be categorized by movement into an even more decentralized government model, one in which the people have a greater degree of access to decision making.

In Switzerland, some citizens have been voting electronically for almost 12 years. Our company, WISeKey, has been on the leading edge of that e-voting initiative. From the inception, we were tasked with helping to create a secure and trusted voting platform, a platform the citizens could rely on. They'd need to be assured that their votes were received, counted, and kept confidential. They'd need to know the will of the people was communicated. And so, we made it our top priority to ensure the voting platform was protected from sophisticated hackers and government manipulation. But how?

In partnership with the government and others, we created a system of encrypted citizen verification. Every potential e-voter was required to visit their post office and present their passport to the appropriate government official as proof of citizenship. Once that passport was validated—much as it'd be when a person enters Switzerland by train or plane—the government official issued a digital identity to the voter. That digital identity could be stored on the voter's mobile phone or on a smart card. It could then be used as a sort of key, in upcoming elections, to unlock the e-voting portal, allowing the voter to transmit their vote across an encrypted and secured government server.

This secure e-voting method has been around for years, but as technology has progressed, we've added additional layers of security. Using WISeKey blockchain technology, we've ensured each vote is recorded on a distributed digital ledger. Much like the application of blockchain technology to currency, this protects against double voting and fraud in vote counting (as votes can be cross-checked against other distributed ledgers) and creates additional barriers to hacking voting systems without physically accessing someone's government-issued digital identity.

Voting systems like the one used by Switzerland provide for a more secure vote and make the process more efficient. Results can be analyzed almost immediately, and electoral winners can be announced within minutes of the polls closing. But it's not all roses and daisies. The system is not without risks. Because votes are recorded with a unique voter identification number, it might be possible to determine how discrete individuals voted. And if that ledger were hacked and voter identities compromised, citizens might be blackmailed, punished, or otherwise disparaged for the ways in which they voted. In this way, without secure protections, the risks of e-voting might have a chilling effect on the polls.

If the challenges to e-voting can be overcome, if people grow accustomed to it, it might help us decentralize our government and allow for the dispersal of power. In governments of the future, we won't simply vote for the best candidate in a two- or three-party system. Instead, because of the ease of e-voting, we'll have more decision-making power. Our representatives will still write and review bills; they may give recommendations and suggestions; and in some cases, where we don't have the same expertise as our representative, we may delegate our votes to him. There may be some situations, though, in which we vote individually, decoupling issues from partisan politics. We'll be able to vote for the things that are best for our families, our neighbors, and our communities.

Through this kind of decentralization, governments of the future will be more granular, and so much of our governance will occur at the level of the state, county, municipality, or even neighborhood. Communities across the globe will vote on issues discrete to them—issues relating to food, water, health care, and utilities. They may also partner with neighboring communities to address specific issues, issues that the national government may not be nimble enough to manage. Together, communities might sort out how to deal with an influx of immigrants or handle material shortages. Communities in Europe could pass local measures to provide resources directly to an African community in the middle of a drought. A local governmental unit in Portland (whether at the city or neighborhood level) might vote to send cryptocurrency directly to a ward in New Orleans that has suffered the effects of a hurricane like Katrina. Instead of flowing through relief organizations and other intermediaries, where the risk of corruption is ever present, funds could go directly to those who need them most, all triggered by a safe and secure community vote. And those local government units receiving the resources can likewise vote on how to best use them—in ways that are more culturally and communally appropriate—all without relying on paternal national governmental structures.

The more granular governments of the future will be connected in other ways, too. Through artificial intelligence, the needs of a community—occupational and commercial; needs for humanitarian services, entertainment, and the like—will be examined in real time, and those needs will be communicated to neighboring communities. The data of individual communities will be monitored, collected, and analyzed in ways that increase citizens' well-being. Organized into smaller civic units, AI will engage in faster and better community modeling, allowing us to enhance our educational systems, reduce poverty, use predictive technologies to better police criminal hotspots, predict future terrorist attacks, and detect cybersecurity attacks. And with faster, more secure means of addressing those problems, we'll create a better, more democratic, more humane government.

Information, Social Media, and the transHuman Politician

In this more decentralized system of government, a system in which citizens have more control over civic outcomes, both citizens and politicians will need better access to more reliable information. The delayed news cycles of yesterday won't do—but the endless streams of manipulative information and "fake news" won't either. We'll need systems that are more reliable and give us the abilities to make more informed political decisions.

In the days before the technological revolution, the days when media companies were more consolidated and slower moving, sensible people from both sides of any argument could at least agree on which news outlets were reliable. There were bounded sets of facts, and only the analysis of those facts was subject to an editorial bent. The golden era of facts has passed us by, though. Now, we live in an era where information and *facts* are two very different things. In the current media ecosystem, and as the Cambridge Analytica case shows, it's become difficult to determine which media sources are reliable and which are peddling misinformation and fake news. And as communication technologies expand, as we're fed more information than we can consume in many, many lifetimes, we'll be more open to manipulation than ever before. If we don't protect our social and information networks from the next Cambridge Analytica, our more granular, more decentralized forms of government will be threatened. And who will benefit the most from these manipulations? Political parties, politicians, and special interest groups.

To the extent that we can protect our social networks from disinformation and fake news attacks (through the right kinds of regulation), they'll be beneficial in the new era of government. Politicians will need these kinds of information networks to spread their messages. So, the political leader in the transHuman era will have their finger on the pulse of social media. They'll use platforms much the way United States President Donald Trump has used Twitter, where he communicates his message, touts his accomplishments, and attacks his opponents. He's used it as a direct line of communication to the people, and whether or not you support his policies or his methods of attack, there can be no doubt he's proven to be among the first of a new breed of politicians. This brand of politics isn't changing anytime soon. In the future, the President Trumps of the world will be the rule, not the exception.

Of course, social media communication won't be a one-way street. Constituents will have more access to their politicians than ever before, and they'll use that access to hold their representatives accountable. Major movements will continue to use those platforms to access politicians too—movements like Black Lives Matter and Brexit and the instigators of the Arab Spring. Politicians of the future will do well to listen to those constituents and movements and engage their concerns before mass protests or revolutions occur. The most popular and trusted politicians will do this well. They'll host social media polls, engage with Facebook and Twitter users, and offer email newsletters that give constituents opportunities to respond directly. Politicians who don't engage on these platforms will find themselves ousted.

Perhaps most importantly, the politician of the future will need to come to terms with his own irrelevance, his own lack of power. He'll need to understand he's no longer as necessary as he once was to the function of democracy, thanks to the increasing decentralization of government functions. Kristin Houser aptly states it this way:

> Thanks to technology, politicians are no longer essential to the formation of an organized society. Initiatives like Democracy Earth, Asgardia, and Artisanopolis are envisioning new forms of society in which the people govern themselves. These societies could have economies that are powered by Bitcoin, governing documents that are drafted through peer-to-peer networks, and decisions that are recorded via blockchains. They needn't apply declarations written centuries ago to today's unique landscape— they can start from the ground up. [10]

As we enter this chaotic era between ages, it will be important for government officials to keep all of these things in mind. Decentralization and efficiency will be the new organizing principles of government, but they'll come with greater regulatory needs. Citizen platforms that allow greater transparency and the spread of more reliable news will be important, and we'll need to protect those networks from disinformation and manipulation. In short, we'll need to ensure that the will of the people rules, not the will of larger, more powerful groups.

Moving Toward a Better, More Liquid Democracy

"Democracy is not an absolute idea. It's a work in progress. It's never finished. It will never be complete." So says Santiago Siri, founder of Democracy Earth, a San Francisco start-up hoping to disrupt democracy. Through promoting what he calls "liquid democracy"—democracy that allows the voter to vote directly on some issues while delegating his vote to representatives on other issues—Siri hopes to bring innovation and transformation to governments across the globe. He thinks that through secure electronic voting, citizens will take a more active role in the decisions that most affect their lives. We agree.[11]

Though technological disruptions are coming to democracy—disruptions like those advanced by Siri—the changes won't happen overnight. They'll come over time, gradually. And as technological advances bring those disruptions, today's governmental systems and politicians will not hold on to power through partisanship or populist rancor. They'll need to adapt. So will we. To start with, we need to make sure we're asking the right questions.

1. Is technological advancement in government making the everyday lives of citizens better, or is it merely bringing more complication, more security risks, and more manipulation?
2. Is technology being used in ways that advance community decision making and connection, or is it being used in ways that support bloated bureaucracies and party entrenchments?

3. Are e-voting systems giving citizens greater access to democracy, or are they creating a chilling effect on the populace?

4. Are networks of information trustworthy and reliable, or are they subject to disinformation, psychographic manipulation, and party-sponsored fake news?

If we answer these questions correctly, the decentralized government of the future will thrive at three levels. At the most granular, most local level, it will empower citizens and allow them to make the decisions that affect their everyday lives. It will also streamline decision making at the national level, allowing the president to operate more like a CEO and less like a supreme ruler. At the global level, it will empower all of us to fight against major global problems. But if we're to harness the power of these more nimble, more efficient forms of self-governance, we'll need to work toward political systems that use modern technology to *empower* the citizenry instead of hold them down. We'll need to elect politicians who aren't afraid to facilitate movement into this new era, instead of fortifying their own positions of power. Rest assured, we'll get there eventually. But can we do this efficiently and peacefully? Only time will tell.

14

INNOVATION

Today's technology is rewiring human ambition. What that means is that technology's timeline is pacing ahead of humanity's. Is this a problem? On one hand, it seems technology has always ridden ahead, in order to introduce the masses to elemental improvements to our lives: fire, electricity, penicillin, and the steam engine. The list goes on. The real question is: Are the aspirations that are being rewritten today actually improving our lives or eroding them? It's always the question humanity asks. But now we must ask it with the knowledge that answering incorrectly will have bigger consequences than, for instance, losing the money we had invested in leech farms or horse-drawn wagons.

Today, we find ourselves in a similar situation to that of the fabled boiling frog. We've been basking in the torrid environment of tech innovation for so long now that we can't feel the burns technology is inflicting upon us. We're watching the bubbles rise while failing to see that we're lounging in a global pot of boiling water. Will we get out before we're cooked? It won't be as easy as it might sound. Contrary to the boiling frog, we might need to rely on indicators other than how the environment around us feels. The advancement feels good right now. It's more convenient. It's fashionable. It's utterly fascinating. Yet, too much good feeling can veil a dangerous undercurrent of human attrition. When it comes to technological innovation, we must be vigilant and honest about where the attrition is occurring. Perhaps recent developments in the surfing industry can provide a thought-provoking illustration of the need for every one of us to consider—

- the nature of our creations (are they fundamentally human?);
- the necessity of them (do they serve humanity first and foremost?); and
- whether they make our experiences on this planet better or worse (as opposed to merely advancing technology in the name of undefined progress).

In an insightful piece for *Outside* magazine, Alex Wilson, the deputy editor of *The Surfer's Journal*, details his experience attending the 2018 Founder's Cup, a professional surfing event held on artificial waves 100 miles inland from the Pacific Ocean. "If you take surfing out of the wild," he wonders, "is it still surfing?" The question is only a surface one. At the heart of Wilson's story is a deeper inquiry into the eventual results of the innovations we undertake and adopt.[1]

Wilson's easterly drive to California's Central Valley takes him through 60 miles of desert, arbitrarily adorned with makeshift signs (hung on abandoned trailers) offering opinions on the state's water crisis. His final stretch is an unassuming two-lane that passes a concrete recycling yard en route to the Surf Ranch, the brainchild of 11-time surfing champion Kelly Slater, which houses, ironically, a 2,000-foot-by-500-foot rectangular pool of aquamarine water capable of producing perfect waves. The artificially perfect waves, only found in nature under ideal conditions, are the manifestation of a lifelong dream for Slater and, according to a veteran surf journalist, the actualization of every surfer's fantasy.

Wilson was underwhelmed at the sight of the pool, which he likened to the irrigation canals he'd passed on the way there. However, he quickly found himself "in awe of the pool's reproducible perfection," which one surfer after another enjoyed, including Slater's fellow world champions Stephanie Gilmore and Mick Fanning. "I obviously wanted to ride it," admits Wilson. "I was even entertained by the contest. After a while, though, I started to feel dried out by the inland heat—then a little bored, with a familiar instinct rising inside me to sneak back to the coast."[2]

Innovation in surfing is nothing new; in truth, it's the primary reason the sport reached global popularity. While surfing's origins can be traced back 1,000 years, to the ancient Polynesian islands, it wasn't until 1777 that Captain James Cook became the first person outside the Polynesian triangle to witness the sport firsthand. It then took almost 200 years more for the sport to reach widespread adoption. Why? Because surfing was seen as a warm-season, warm-climate activity relegated to places like California, Australia, and Hawaii, each becoming synonymous with the sport and lifestyle. It wasn't until a former Navy Air Corps pilot named Jack O'Neill began experimenting with a synthetic rubber material called neoprene, in the late 1950s, that the perceived boundaries of surfing were expanded.[3,4]

Officially, history credits a Berkeley physicist named Hugh Bradner with inventing the wetsuit, but it was Jack O'Neill who was savvy enough to iterate and market the item into the industry's consciousness. By the mid 1960s, wetsuits were outselling surfboards, enabling surfers to paddle out in winter, when the waves were usually the biggest. From there, the sport's expansion took off north, up the coastlines of the U.S., where ocean temperatures had previously been prohibitive—and then east, to Europe, and west, from Australia to Asia. Innovation through technology has since remained a driving force of the surfing industry, making more waves accessible and producing cheaper, more effective equipment. In many ways, surfing has become one of the purest microcosms of humanity's relationship with today's innovation.[5,6]

On one hand, without technological progress, surfing would still be relegated to warm coastal climates. While there are some who wish that were still the case—although the number is quite small now—the majority who call surfing a favorite pastime embrace innovation as the sport's way forward, even knowing that certain upgrades result in more crowded breaks. One of surfing's greatest allures remains the untamed nature of the activity and the mystery of undiscovered, yet-to-be-ridden waves. In this sense, innovation is a critical bridge to exploration and the spoils thereof. Laird Hamilton is widely known as the greatest big wave surfer in history, not only for his inimitable skill but also for introducing the sport to the concept of tow-in surfing, a technique whereby a surfer is towed into a large wave by a motorized vehicle, usually a Jet Ski. This technique allows a surfer to "catch" waves that are effectively uncatchable when paddling by hand, those approximately 30 feet and up. What's interesting is that when Hamilton first brought the technique to his sport, a massive wave of backlash rose up from within the industry. Many felt it was cheating—that is was not really surfing if you didn't paddle into the wave under your own power. There was, and still is, validation for this feeling. Surfing's origins involve one Hawaiian rider standing atop a specially shaped plank of koa wood and sliding down a breaking ocean wave. That original method of surfing is still possible today; the reality, however, is that few do it that way anymore. Recall the allure previously mentioned: the untamed nature of the sport and the mystery of yet-ridden waves. Allowing that allure to move you almost necessitates an allowance for innovation.[7]

Where would surfing be today if not for fiberglass, fins, or neoprene? Every major development just mentioned has been met with waves of backlash. But eventually they subsided, like a storm's swell always does. Does this mean every major innovation can and should be adopted? Absolutely not. Nor does it mean that adoption must apply to everyone in the same manner.

Bethany Hamilton (no relation to Laird) is a fascinating example of a surfer for whom modern technology seems a perfect fit. If you've seen the 2011 film *Soul Surfer*, you're familiar with the tragic story of Hamilton (then 13 years old) losing her left arm to a tiger shark while surfing. After months of recovery, and a longer season of coming to terms with her loss, she took up surfing again. Many assumed a natural first step was to wear a prosthetic arm to help her pop up and maintain her balance on a board. It wasn't the route she chose.

Instead of embracing technological innovation to do what she loved, she innovated herself—her psychology, her physical strength, and her abilities. In other words, she already had what she needed within her. It's an option worth considering in every realm and industry that innovation touches.[8]

There are three overarching questions that should be addressed when it comes to innovation:

1. What can be artificially enhanced or re-created? This is a question of technological resources and capabilities.
2. What should be artificially enhanced or re-created? This is a question of ethics, morals, and, ultimately, personal choice.

3. What will be artificially enhanced or re-created? This is also a question of personal choice, but in this context, personal choice is often shaped and swayed by societal norms, consumer resources, and government policy.

What we do with our current ability to innovate can be summed up this way: Just because we *can* doesn't mean we *should*. Just because we *should* doesn't mean we *will*. And just because we *will* doesn't mean we *can continue*. There is no perfect approach for innovation today, except to accept that it is both personal and pervasive. We have an obligation to consider the implications of both applications. Perhaps we should start by considering whether everything can be improved in the first place. We have a natural world still ripe with elements as close to architectural and technological perfection as we've ever seen. We have bodies and minds so fascinating and technically complex that we understand only a fraction of how they work. Are there pieces of our world—both outside our physical bodies and inside them—for which innovation has no use?

Iceland is one of the biggest islands in the northeast Atlantic, with nearly 3,000 miles of shoreline and frigid water temperatures hovering around 39 degrees Fahrenheit in winter. It's the last place you'd think of for a surfing trip. But the waves (not to mention the entire landscape) are said to be pristine and barren in December. For this reason, renowned photographer and longtime waterman Chris Burkard teamed up with several cold-water surfers, including two locals, to document an epic surf trip to Hornstrandir National Park, located on the northernmost coast of the island.

"We finally arrived at the national park by boat, and the surfers (in thick seven-millimeter neoprene wetsuits) started paddling out into frigid waters," Burkard said regarding the genesis of the documentary, aptly titled "Under an Arctic Sky." "Then the boat captain told us a storm was approaching—and fast. We reluctantly turned the boat around and headed back to the harbor. I felt super guilty as I was the one who promised adventure, epic waves, and I put their lives on the line. Ultimately I felt I let everyone down. It was overwhelming."[9]

Monitoring the weather, the crew saw this wasn't just a snowstorm—it was a full-blown tempest named Diddu. A storm larger than anything Iceland had seen in the last 25 years, weather predictions put maximum wind speeds at an astounding 160 miles per hour, along with the potential for level 4 avalanches. Completely disheartened, and at the risk of disappointing sponsors and spending more money, the crew decided to leave—until an odd feeling told them not to give up.

"Even though our decision to ride out the storm probably wasn't the safest idea, we also realized that as the conditions grew worse, it brought the most incredible swells we've ever seen," Burkard recalled.[10]

After 18 hours of treacherous driving on the edges of cliffs in total darkness, including digging their trucks out of road slides, they finally became stranded in a cabin near the coastline. Despite mounting exhaustion and disappointment, the crew couldn't stand to stay cooped up in the small home. Just as they walked outside, the storm broke. What happened next was one for the history books.

"These swirls of neon green, orange, red, and yellow light started to appear—it was the Northern Lights," Burkard remembered. "Then the moon came out and the Northern Lights grew more intense in color. I can't even describe the amount of luck that went into this or the transcendence of that moment. We grabbed our gear, got the surfers in the water and started shooting. We were all caught between the overwhelming beauty surrounding us and trying to remain focused and professional."[11]

Watching the final scene that Burkard describes is a sublime experience, the sort of awe-inspiring encounter with beauty and mystery that couldn't be felt other than through an interaction with the natural world. The truth, however, is that the encounter was only natural for Burkard and his surfing comrades. For those of us who watch the film, the viewing experience is actually occurring through a technological innovation called HD video, shown on an HD screen that uses liquid crystals to reflect the images Burkard and his friends witnessed firsthand. However, to doubt the compelling effect of watching Burkard and his friends surf Icelandic waves under the Northern Lights is missing the point. It is an absolutely compelling viewing experience, and we have technological innovation to thank. But we must also acknowledge the role that the natural world still plays.

Without technological innovation reproducing their experiences, doubt still lingers around the initial claims of early explorers like Americans Frederick Cook and Robert Peary. In 1908 and 1909, respectively, each believed he was the first to reach the North Pole, but we have only handwritten journal entries and a few photos of heavily clothed men standing in a snow-covered landscape. Whether they actually made it remains to be seen, as does a more visceral representation of what it was like as they explored the northernmost regions of earth. A photo and a firsthand account help our imagination construct a fuzzy still frame—but it's not HD video. The fuzzy frame won't move us the way Burkard's documentary does.[12]

Some say the Snapping and vlogging and documenting of experiences today is a shame. They say it's taken the wonder, even the desire, out of experiencing life firsthand. We should get out more, they say. Be present more. Turn off our phones and cameras and see the world with our eyes— feel it with our senses.

Perhaps those who say such things have a point. There is nothing quite like a firsthand experience. But what if you're bound to a wheelchair, or bound by financial constraints, or simply bound by the confines of the human experience? We thought the earth was flat at one time. We thought cars were a pipe dream. And we thought there wouldn't be a market for personal computers. What are we ignorant about today? Plenty, which is why the future is so exciting.

How can we pursue this adventure alongside technology? It starts with an acknowledgment that innovation is both an internal and an external process. In other words, it comes from someplace inside our hearts, minds, and souls that we can't fully explain. Who really comprehends the birth of an idea? No one—we just know it comes from somewhere inside of us. But innovation also comes from the outside, through the external context in which we find ourselves: members of a larger, equally incomprehensible universe. What can we learn from a molecule, a maple leaf, a moon, or a Pacific sunset? What do the mighty coastal waves say about motion and leverage—and even life itself?

One of the great mysteries of humanity is that innovation comes from inside of us and outside of us at the same time. We shouldn't shy away from the mystery but embrace it. We should stay inside and get out. Look inward and gaze outward. The lessons we need are flowing inside of us and floating all around us.

In his book *Crazy for the Storm*, author Norman Ollestad touches on this amazing dynamic of human existence when he describes catching what surfers call a "tube" for the first time as a young boy, during a trip to Mexico with his father.[13] When the ride ends, Ollestad's father paddles to him. "You've been to a place that very few people in this world have ever gone," he tells his son. Later, while sitting around a fire with some local villagers, a teenage girl offers a more concise description: It is a "doorway to heaven," she says. Ollestad recounts staring at the flames rising from the fire and thinking, "Beautiful things were sometimes mixed up with treacherous things, they could even happen at the same time, or one could lead to the other."[14]

He could very well be talking about today's convergence of technology and humanity.

Technology has long shaped the world we live in. From fire pits to wheels to electricity and the computerization of just about everything. But human innovation ultimately shapes technology and its applications. Perhaps with the only exception being our ability to love one another, when, where, why, and how we innovate comprise the most fundamental human force in the universe—and innovation can even improve how we love. "We are born makers," writes Brené Brown. "We move what we're learning from our heads to our hearts through our hands."[15]

Creating is in our DNA. Resourcefulness is our single greatest resource. Today, the right application of this intrinsic act is more important than ever. We must give thought to what we create and why we create it. Is the experience we're after better for humanity, or just a temporary hit of adrenaline that will detract from our appreciation of something the natural world has already given us? Will our creation improve our unmatched natural capabilities or diminish them?

We must also understand that if we do not create, we are doing ourselves, and humanity at large, an enormous disservice. Ultimately, the future will not come down to a competition between humans and machines. We are the creators. Will we aim to unleash the greater capacity that's already within us? Or will we overlook our untapped capacity for an easier alternative? The real competition to determine the future is the battle we will have with ourselves. Let's ensure our best selves win—so that we win together.

THE FIRST STEPS TO A HUMAN-CENTRIC FUTURE

The objective of this book has been to highlight the future of humanity along the continuum between a human-controlled world and a machine-controlled world. The major assumption of the book is that we should start from somewhere in order to define this new world—but that there is great risk in starting this design, let alone this discussion, from the machine perspective. The transHuman code must begin and end with human input, human emotion, human ingenuity, and collaboration, basically with humanity as the center of gravity of the future society we are creating.

We have also set out to establish here the foundation of a human-first platform: a place online where we can continue important conversations and collaborations, define standards, involve multiple stakeholders, and define a philosophy we collectively ascribe to. From this platform, we can ensure that the beautiful, transcendent complexities of humanity can be strategically and safely incorporated into the technological pillars of the future world.

The theme of the 2019 World Economic Forum in Davos, Switzerland, is "Globalization 4.0: Shaping a New Architecture in the Age of the Fourth Industrial Revolution." It seems only fitting, as the transHuman code initiative was conceived here—in a conversation between us and the World Wide Web founder Tim Berners-Lee, in January 2017—that we return here for its formal introduction.

In his introduction to this year's agenda for the annual gathering of global business, academic, religious, philanthropic, and civic leaders, titled, "Globalization 4.0—what it means and how it could benefit us all," Mr. Klaus Schwab, the WEF's founder and executive, asserts that our global governance architecture is under fire. We acknowledge that the Fourth Industrial Revolution is coinciding with dramatic changes in governance, realignment of international order, and increasing disparity between the haves and have-nots.

The growing economic divide is attributable in large part to the impact that the digital transformation is having on our economic foundation. Our value system is being reset as data rises to the top of the list of most valued assets. It is Mr. Schwab's premise, like ours here, that the unprecedented change to the elements of our life ecosystem—health, transportation, and communication, to name a few—warrants new frameworks for government and corporate collaboration and a new approach to education. Mr. Schwab tells us, "Globalization 4.0 has only just begun, but we are already vastly underprepared for it. Clinging to an outdated mindset and tinkering with our existing processes and institutions will not do. Rather, we need to redesign them from the ground up, so that we can capitalize on the new opportunities that await us, while avoiding the kind of disruptions that we are witnessing today."

The core premise of the transHuman code platform is a multibillion-stakeholder approach to the invention of our future. We simply ask this:

What if we all agreed to the meeting of our most basic, common, and critical needs? And what if we agreed to global accountability to this end? It is possible and, in many ways, quite necessary.

With enough collective intelligence and effort, this platform will, like an artificial intelligence–powered operating system, not only power the world we desire for humans but also guard against anything hostile to it. In doing so, it will act much like our own immune systems, able to detect a virus and immediately set out to attack it.

While it is also possible that this human-centric future we imagine will occur without global consensus, history suggests that it won't. It is time we humble ourselves and agree that no one constituent, industry, or country has the perfect answer. There are enemies of humanity—some of whom don't see themselves as such. There are also many self-interests posing as friends of humanity. These are difficult to spot and avoid if we are divided; they are not difficult to spot if we share the same vision. Every human matters today. Not only from a humanitarian standpoint, either, but also from a practical standpoint.

Seventy years ago, the United Nations General Assembly adopted the Universal Declaration of Human Rights. The Declaration was a groundbreaking agreement affirming the rights of individual citizens, humans, including the right to freedom from discrimination, the right to education, the right to a free and fair world, and many more.

The Fourth Industrial Revolution is anchored in technological innovation, changing the way we live, work, and interact with one another. It also has the potential to both challenge and uphold human rights and challenge humanity all together.

Technology will define us, instead of the other way around, if we allow fragmented innovation and adoption. That's not to say we advocate a universal system of laws to constrain us. It is to say that unless (at least) the majority of us put humanity first and assert ourselves in this cause, another majority will arise, as it always has. And it generally forms by way of either force or finance. These cannot weigh down the future we desire, let alone derail it altogether.

It's no longer enough to assert that democracy is supreme. Democracy must be shaped by us, through the one global relationship that can either destroy it or elevate it indefinitely: humanity's partnership with technology. How we lead, or don't lead, this partnership over these next few years will define life on this planet indefinitely. Let's lead. All of us. Together.

CONVERSATIONS WITH
THE INNOVATORS

In the creation of this book, we have had the privilege of reading, listening to, and engaging in conversation with many dynamic innovators in the world today. On the pages that follow, we want to share firsthand the inspiration, perspective, and action of a few of the individuals who are shaping our future.

Our hope in sharing them is to ignite further innovation and inspire a greater level of collaboration for good the world over.

Conversation with Kavita Gupta

Founding managing partner at ConsenSys Ventures, social finance pioneer, and leading innovator in technology investment and advancement

Authors: Kavita, you are truly a celebrated innovator for social advancement, an emerging markets technology investment trailblazer, and now a premier venture capital investor for blockchain.

As background, the transHuman code project began as a roundtable conversation two years ago at Davos, with industry pioneers Tim Berners-Lee, the creator of the World Wide Web, and Wikipedia founder Jimmy Wales, about technology and trust. This came at a time when concern was growing over the Internet platform companies and people's understanding of what they own and what they don't own.

Gupta: And now we all know about it, but we still use them, don't we? We don't live in a perfect world!

Authors: Our question was, "What should the relationship look like between individuals and technology solutions in the absence of a global governor of technology?" Since there isn't anybody that says, "Let's do that; let's not do that," how do the developers of technology, the users, and the promoters of technology align so that we're actually getting more good done for humanity?

We believe that the best way to do that is to continue a dialogue around what is good and what can be accomplished, trying to keep at the forefront of the conversation that the goal is that more people will have a better life, as opposed to a few people.

As the leader of your organization, as you're exploring new products and services in that application, it is our recommendation that you'd be pushing everything through a filter, which is the transHuman code—a code that represents the human traits and values we collectively want to see flourish more than they are now.

Gupta: Yes, I absolutely agree. This in fact was a focus of a conversation I had just yesterday. And again, just a week ago, I was speaking on the human values of new technology at Oxford. I'm actually now writing on the importance of establishing, as core criteria, the ideal life balance for any app or any technology. This would be sent to our developers before they start on the big update or start getting it in people's hands. But without creating a layer of rules to hinder innovation. We don't want it to become an automobile industry where you can't do innovation because you have to have so many licenses, so you input one. We don't want to overregulate, but we also need to make sure companies aren't becoming too strong, like Lime Scooter, for example. It's dangerous. It's super, super dangerous.

Authors: It didn't put people first, did it?

Gupta: It didn't. Forget about it. Neither the driver nor the people on the street. They got around the regulations and now, apart from wearing a helmet, there's literally no rules.

Authors: It's commercially driven, right?

Gupta: Exactly. And it's fine as an industry for things to be commercially driven. We talk about minimum viable product, but maybe we should also be considering what is the minimum viable consumer good? Ideally, you should have to do good to make things available.

Authors: So who's going to be the governor? Collectively it's difficult, right? But leaders like you and leadership organizations like yours have the ability to impart that message and influence, not only within your own organization but also with other organizations that look up to you.

Gupta: Yes, I think as an individual, definitely. It's more difficult on the corporate level, especially when it comes to engineering. As a student, following the engineering path in college, there was not even a single policy draft that was talked about, "Hey, how do you think about this project humanely, and what are the things you have to prevent or promote?"

I think that conversation should be engaged in by the software engineers. And it should start at a very early stage, much like it does for the person becoming a doctor, so that when the engineer eventually has a corporate leader say, "Hey, let's create this," the engineer is automatically saying back, "Yes, but does it address this basic human principle? How do I make sure I create a product which is improving the life for most people?"

So I think the transHuman code should take that path. It is innate in us that this is our responsibility, but we need to keep that at the forefront of our training as engineers.

Authors: This is a message that the whole world will benefit from when it's conveyed from people like yourself, that are innovating.

The medical school analogy is really good. That would be a good message to reinforce. Ethics is a key element. Neither of us have been to medical school, but we know that it's a key element that's taught in medical school, the same way that it is in law school.

Gupta: Yes, in medical school—my brother's a doctor—I remember every semester they have an ethics class. Every semester. And nowadays, once they define their specialization, they increase the ethics classes according to that specialization. But we don't have that in computer science classes anywhere, at any point.

Authors: It wasn't part of our business school regimen either.

Let's switch gears a little. We're interested in hearing your perspective from the investment side of your experience, especially where the spotting of good innovation is concerned. What common qualities, the most important qualities, do you look for in the innovators that you invest in and/or work with?

Gupta: There are two key qualities we look for. The first is their awareness of the whole picture. So, if you're looking at the founders, what has been their background, how well do they know their product or the need of that product? And when they're thinking about that need of the product, have they done thorough research on that market? And what is their connection to that market? And then on the technology side, will they be able to deliver or get the team together to deliver it? And together what could the result be if you align these capabilities?

The second thing is how they think about the customers, are they thinking about their customers purely from the financial perspective? That will not only tell you how they're going to target their audience, it will also tell you how they're going to run the company.

When an entrepreneur tells me, "This is what I'm developing," I say, "Do we really need this in the market?" I start there. Then I move to, "What is the competition?" So let's say, for example, you're building a mobile wallet for crypto. Definitely there's a lot of competition, but there are still gaps in the space. "So how are you different from what everybody is providing?" They could be different because they have a very good usability, UX/UI team, who understands how to make a seamless integration for paying customers. Or they could be a very strong blockchain team and say, "This is for the existing customers who understand how to use Crypto, and we're just going to make an even more secure wallet." And then once you've narrowed down who you're targeting within the existing market, I want to know, with respect to the competition, how do they stand out? Once you understand how you stand out, is the team now strong enough to support that position or not? So the criteria basically starts from the product gap to the team, to operations, to customers adopting the product, and then to the business. I don't start with the business because the business model can change. And that is the least risky thing to change at the last minute. But it's more difficult to suddenly change your team, change your target audience, change your product.

Authors: Do you find that it is difficult for some entrepreneurs to embrace flexibility with regard to the direction that they want to take the business? You talked about the differentiation; what is the differentiation in the market? Are they usually open to better options? And is that a fight that you have to fight often, or do you find that most entrepreneurs are pretty flexible?

Gupta: A lot of them are flexible, but sometimes you, as the VC, can't see something that they do, either. So, the flexibility must go both ways. There has been a case in my own experience where I passed on an investment, and the founder ended up creating a billion-dollar start-up. They led the company in the direction they wanted to go, and I couldn't see it working, and I just kept on saying, "Guys, I don't think it's going to work. Who's going to do it? Somebody will kill me if I invest in this." It seemed so insecure, and they ended up creating a billion-dollar market, and I use their product now, every day. It taught me, early in my career, that I am going to come across some super geniuses or super lucky people who would have the vision that I don't have and the capacity to understand how it will work. But you do deal with certain personalities who just keep on pushing it. At some point you just say, "Hey, I really believed in your earlier product," or, "I really believe in you." And then you tell them it's fine if you want to go and experiment. Hopefully, it won't be too late for you to get it if something works. And you leave it there.

Authors: Our observation of entrepreneurs is that there are three characteristics that are inherent. There are several different schools of thought on when these characteristics evolved, but nonetheless we consistently see them as creativity, perseverance, and tenacity. These can be expanded upon, and they can be fed throughout the individual's life. But flexibility, in contrast, is most often a learned skill. And it comes from rejection more often than anything else. And the same thing would be applicable to technological innovators. So that billion-dollar opportunity that you missed . . . you may not have been the one to impart the change necessary for it to occur, but someone else was. And/or the frequency of your messaging might have had an impact on the founders but not until after you moved on. But ultimately, you don't want them to be less persistent, less tenacious, less creative. However, if they're looking for money, if they're looking for capital, they're going to have to fit into your box somehow, right?

Gupta: I completely agree, but I can be perseverant and tenacious for a particular part of my personality, and completely oppose it for other things. Many entrepreneurs are flexible once they receive feedback, and that's what the successful combination is. If you're going to be persistent and tenacious on everything, then you will never be able to create something that makes perfect sense to people. But going back to the original question, I do see that the pushback comes from extreme cases and from different thought process. And as an investor, everyone knows you can't push back on the original thought process, because there is not usually a place for you there.

Authors: Thank you for that insightful explanation. Let's take on a new topic that you just brought up: blockchain. For the average reader of this particular book, or average consumer who takes part in this platform that we're creating, blockchain is a term that he or she will know, because of the popularity of it; however, I think their understanding of the concept will have some gaps. How are you envisioning and describing blockchain to the average consumer today?

Gupta: I actually end up doing it in a lot of situations, especially with my mom and dad. They say, "We want to brag about the work you do, but we don't understand what you do." [Laughs.] This is how I usually speak, if somebody's sitting next to me and they're like, "Okay, so what the hell is this blockchain? Is this Bitcoin or something?" I usually explain that, first, blockchain is something that gives you complete data privacy, because your data is moving with your permission. And I explain that instead of getting a notification that says your work email account or your Gmail account was hacked, because the provider put all your data in one place, what if all your data is separated into such small pieces, around a thousand nodes. Now, if the hacker gets into the first node, the other node is down, and all they've got is one of them. And they can't put it together without some serious work. They'd have to really be after you to spend the time to put that all together. And so that's the data privacy and data security side. Then there's data monetization, which always puts a smile on people's faces. I explain that over time we're going to start seeing that every time you agree to give away your data, you receive a monetary incentive like a rebate or discount. So, for those not working with blockchain in their business activities, your data is no longer freely available, and you control how it's used and who ultimately gets to use it.

Authors: Can you give us your take on the role that capital plays in innovation? ConsenSys established a $50 million fund. Why $50 million? And what are your priorities with $50 million? How can you best enable innovators? And what does innovation success look like for you and ConsenSys?

Gupta: I think that's really a paradoxical question. And that's in my mind every time I'm taking an investment. Capital can definitely fuel completely new types of innovation. I was one of the first people from ISU who landed in Nairobi in 2006 and said, "Oh, Africa has amazing tech innovation going on. They invested in a payment system like Pay-Pal, and we should invest in it." But I still remember my VP, an amazing woman, the first ever woman to hold that investment role at IHC or World Bank, said, "They don't have food to eat; what are you talking about?" I said, "Everyone has a cell phone here. They don't have food? I don't know why they don't have food, but they have cell phones." I pushed and pushed, and we created the first ever investment in the fintech field in Africa. And that capital enabled entrepreneurs and the market to evolve, and I know we were influencers in the creation of a multibillion-dollar private economy. So, this is one of the major benefits if you start moving capital into a particular industry where others are not yet but innovation is possible. You can create new jobs, you can create new industry, you can create new entrepreneurs. For me personally, supporting entrepreneurs around the world is so important. Really believing in them and their products and saying, "I don't care what college they come from, I don't care what geography they come from, and I don't care what sort of background they come from. If they can deliver something good, I'm going to invest in it."

Venture capital investment in innovation is supporting a whole new geographic generation, and that stake is paying in more ways than just financially. We were one of the first to invest into new companies in Australia, in Chile, in Egypt, in Poland that came out of nowhere. And then suddenly a lot of US investors started financing them. Rather suddenly, in the blockchain space, it doesn't matter whether you're from Cincinnati or Nairobi, you can really attract money if you have a good blockchain product that people will benefit from.

So as a capital investor, you can change the norms, you can support innovation in a different way. It will create new standards. Now, the negative part of it, the downside of it is that it takes time. You do end up burning through a couple of investments to really start making progress. But through capital you can start your own revolution. For me, creating African markets was a revolution, like saying geography has no stake in investment, or saying I don't care which college you went to. Investment capital is making inroads to the places that traditional money won't go. What does success look like? What parameters do you measure it with? If it's financially? Then yes, three X, four X return is a success. If you're thinking philosophically, for me the people who were sitting on great innovations but were missing either an ecosystem or that next exposure, they got it because of our capital, and they ended up creating amazing product. That's a personal success to me.

And at the end of the day, you have to feel comfortable that these people are going to treat other people very well, that they don't end up being assholes, obviously, while they create an amazing team, and see people working, and you see that they value that human interaction or human values while they are creating multimillion dollar companies, and they're not cheating their customers.

We created an accelerator at ConsenSys, and very early on we started talking about ethics with our customers. "How do we create the right dialogue, that we are not shortchanging our business, but we are also not shortchanging our customers. How do we be sincere and transparent, that we are here to give you a product, but the product is not going to take advantage of you. So start by building that protection there on our data transparency layer."

Authors: You really embody the core of the transHuman code message. You're truly inspired to enable other people regardless of what their background is. But you clearly want them to do good things for humanity in the process, as a result of your investment, personally and financially.

Gupta: Yes. So often, in the first 10 years of a business development you're just trying to be successful on paper with respect to returns, with respect to what did you do. After you have achieved that, we all somehow with our maturity start thinking about what we can do for society. But why don't we just start on that at the same time?

I would be remiss in this transHuman code discussion if I did not introduce a subject that I feel is so important today. We have an amazing global conversation on gender going around right now, which is great. But it doesn't mean to me that I should be investing in a start-up because it has women. This is one thing that is also very close to my heart, and I think I don't have the popular answer to it, but I have an honest answer. I bring the same philosophy to investment opportunities that I do to my own career. I don't want to get any position if I'm there just because I'm a woman. Because that means everything I've done has been shortchanged for the sake of elevating where I was born, my sex, or what I look like. Inclusivity should mean that it doesn't matter who you are . . . if you are good at something, you are given the same opportunity as anybody else who's good at that same something. That's what true inclusivity is.

Authors: This conversation is both inspiring and encouraging, Kavita. It reinforces the core belief of the transHuman code that good can prevail during this technological revolution through the stewardship of wise and dedicated people like you.

Gupta: Thank you. ▲

Conversation with Alex "Sandy" Pentland

Cofounder of the MIT Media Lab, Toshiba professor at MIT, serial entrepreneur, and one of the most cited authors in computer science

Authors: Sandy, you are a renowned thought leader, an esteemed educator, and a serial entrepreneur. And in each of these endeavors, innovation prevails. In preparation for this discussion today, we are reminded of our "Gathering of Minds" discussion on the transHuman code at Davos last year and how your optimism on the future of humanity and technology was both inspiring and encouraging in the face of gloomy pessimism being expressed by others.

Pentland: Well, if I look at most of the world and most of history, life is hardly fair or democratic, and it's only a small layer of members of the international culture that are able to live what we would think of as the life that we'd like to live. But that small layer has grown tremendously, factors of 10 and even 100, in the last century.

So, I look at the arc of it and say, well, first of all, there's more to be done than people appreciate. You have to actually live in a place like India, South America, Africa to appreciate that the life that we see is not the life that others in these countries live.

For instance, I remember when I would go to different cities where I was setting up laboratories at universities, and everybody knew everybody. What that told me was that there was a group of maybe 10,000 who were that international class in all media, and they ran everything, but 10,000 people is nothing in a population with a billion. And the same thing if you go to Davos, 3,000 people, and through those people you can reach almost everybody in the international class. What that tells me is there's something like maybe a million people, say, that are members of this international culture that has true opportunity. But most people don't.

Well, when I started being really involved in international development, the common way you would start a talk was, "Half the people in the world have never made a phone call," and today, 90 to 95 percent of all adults in the entire human race own a phone, a digital two-way communication device, and in a couple years, the vast majority of those, the vast majority will be smartphones, which means they have something approximating Internet. Now, you can be negative and say, "Oh, it's not free Internet," but compared to what it was even 10 years ago, it's unbelievable.

Similar, people go, "Oh, AI. AI's going to take over the world." Well, all of those people with those phones walk around with four or five AI applications in there that were considered century-ahead science fiction only 20 to 30 years ago. If you go look at, say, *Star Trek*, the movies or the television show, and the idea of having a computer understand you in your natural language, the idea of translation, you have that on your phone today. It's incredible. In *Star Trek*, it was seen as something for centuries in the future, but we're all completely used to it across the whole population of humanity. Maps that let you know where everything is, that was inconceivable, I mean, absolutely inconceivable only 30 years ago, and on and on and on. Things like that which were future dreams have now become real, and we use them and we're happy with them. We're not going to give them up. We are in love with them.

And, yeah, there are problems with them, there are worries about them, but we've come so far that we hardly, hardly remember what it was like before you had facilities like Google, before you had maps, before you had communication with anybody you want. So, I think it's very difficult . . . it's a serious mistake not to be optimistic. We are aware of the dangers, we're discovering the dangers, we're anticipating them, we're doing things in response. There are both governance solutions and technical solutions for most of these things.

In fact, the single biggest thing to be pessimistic about is human organization, the way we do politics, the way we make decisions, the way we run companies. That is perhaps the one important thing in our environment that hasn't changed. We still do business pretty much the way we did centuries ago. That, however, is finally beginning to change, and the thing that's changing it is the fact that there is now data about everything.

So, the census is no longer how many people live in a place, it's also what information . . . how much money do they make, what sickness do they get, how much are they educated? The data that we know about ourselves is getting richer and richer, and we're at an inflection where we'll become dramatically richer to know ourselves. In fact, that's one of the grand achievements with the sustainable goals is to write down 169 things that every country should measure continuously about its population that has to do with sustainability, that has to do with inequality, you name it, and as that data becomes public, it simply becomes a regular process to be able to publish that data, and it will change the character of what we're able to achieve. It's a transformative transparency and transformative accountability that is upon us and we're hardly aware of it, but yet we already have the tools in place.

Excuse me for going on about this, but I just get perplexed. For instance, take global warming. The only reason we know about global warming is these data resources that we're seeing as threats to privacy, things like satellites and satellite imagery—things like some of the transportation and other infrastructure data technologies. We know about these things because of these new data resources, and, all of a sudden, we discover there are problems, and as we discover there are problems, we can begin to fix them. The primary barrier, again, is we have to learn how to have a better, more accepted discussion among ourselves.

So, the part of the transHuman Code that I like is the ability to coordinate with other people around actual facts, data, things that seem to be objectively true, and I believe that the more we can do that, the more effective we'll be as a species, and the more we should be optimistic.

Authors: It's common folklore that suggests that the most valuable innovations are born from individuals working alone in obscurity. The suggestion is these mad scientists or geniuses create solutions in their labs and bless the world with it when the time comes. We know that's not the case, so overlooking these popular caricatures—can you describe the importance of the intersection between innovation and multiperson and multidisciplinary collaboration?

Pentland: It's interesting to know where the story of the individual mad scientist came from. My guess . . . I know the historical facts, but my guess is that it is actually a political statement. It's a political statement because into the 1700s, truth came from the king and the pope. Individuals had no say at all, and this notion that individual intelligence, individual rationality, individual decision making, individual creativity . . . it all countered the king and the pope. And yet we eventually learned that it is this very thing that transforms societies to be democratic, free. That story of the individual has been the most transformative thing in, perhaps, the last millennium or two—whether we're talking about changing how electricity is used or changing a society.

However, it's not a very true story. All the data that's collected on people shows that at least half of our behavior is due to learning from other people, and that's where fads, trends, and norms come from, and the more data we get, the more insight we get into actual behavior, the more we see that's true.

The same is true for innovation and creativity. Einstein did not work alone. In fact, the best analyses of Einstein say that his discovery was a slight twist on what Maxwell did 40 years earlier. And Maxwell himself was a twist on earlier things, and it wasn't just this one thread, it was a whole fabric of discussions and interactions, and that's what we see in our data. The thing that ultimately drives innovation is different cultures interacting. It's not necessarily diversity in terms of gender or race or things like that, though that can certainly help; it's really driven by different cultures interacting. Now, gender and race have somewhat different cultures, but when we look at cities around the world, we find that we can predict rates of innovation. Not just patents, but increase in GDP. You can also predict the negative, the opposite of that, crime and class, by the amount of culture diversity that is or is not there.

One of my students just did a study of all of the young companies in China through the national incubator system, so 5,000 young companies. And they looked at everything they could think of, and the single biggest factor in success of young companies across all of China was cultural diversity. "Had these people had experience outside of the standard culture of China?" The second biggest factor was technical diversity. "Had they worked in different industries?" So, that's astounding, that across perhaps the most innovative, rapidly changing society on earth, it's diversity that is by far the greatest influencer. And this is by a factor of two, three, four larger than the things we normally think of . . . diversity is the predictor.

Similarly, when we look at cities and districts and neighbors in the Far

East, in Europe, and in the US, if you want the biggest predictor of conditions there, it has to do with diversity in terms of communities coming together. That accounts for almost half of your success of your variants in terms of innovation and what you would call progress, the improvement of conditions.

So, that's a very different story than the rational individual story. It's a story of social fabric rather than an individual. And the truth is, from all we know about biology and psychology is that humans are both of those things. Our ancestry is . . . and this is preindustrial . . . is, as a social species, the social fabric is dominant. And we developed language, we developed the individual characteristics, but the social fabric is still half of everything, and the way we manage ourselves tends to ignore that. We talk about individuals, not communities. We talk about genius, not innovative cultures. We talk about personal growth far more than collective growth.

Authors: What's fascinating about what you're saying is that it makes us think of the phrase "Jack of all trades, master of none." Are you aware of the origin of that phrase, where that came from originally?

Pentland: No, I don't know where that came from. Where did it come from?

Authors: That phrase was penned roughly 500 years ago by a British dramatist named Robert Greene, who nobody's heard of. He was describing a young actor and dramatist who didn't seem to be able to find one particular thing he liked doing and so he pursued several creative endeavors. The name of that young guy he was describing as a master of none was William Shakespeare.

This idea that you're describing, of being collaborative, being open, not being individualistic, being open to opportunity anywhere, diverse cultures, plays right into that idea that perhaps maybe we ought not to be so focused on mastery.

Pentland: I think that's exactly right. And also, if you think about it, individuals such as Leonardo da Vinci, they're often described as Renaissance people, and I think it wasn't just that they were really smart people. You could tell the story that, "Oh, they're just this polymath genius," but I think actually they were much more like the jack of all trades, exploring varying disciplines at all times.

Authors: Right. So, in that vein, how do we begin to think more innovatively?

Pentland: Well, I can tell you what we do here in my group, in my laboratory. The MIT Media Lab is famous for being a center of innovation, and what I do has a pretty reliable track record, and it really has to do with the fact that this is one of the few places in the world where industry, academia, and civic systems come together. Everybody in the world comes through here. The reason I go to Davos year after year is exactly the same; it's the one place industry, government, and civic institutions come together, and from all over the world.

So, the Media Lab has people from all over the world. They have people with diverse backgrounds, both those who study and those who contribute financially, they are from all walks of life, all types of industry around the world. And the key mechanism that we have here for research is that everybody has to explain their research to all these different communities. You have to be able to say what your work has to do with civic systems in the Middle East, with industry in Africa, with academia in Japan.

And if you can explain why what you're doing is important to all those communities, then you're doing something that's genuinely more important than you might ever realize. You've actually gotten something that counts as a critical or important innovation. And I think that's the key thing across these communities, creating the biggest opportunity for impact. And then pursue that across the different communities. I think that is where the best changes, innovations, and impacts come from.

Authors: The idea of the main criterion being that you have to articulate the concept and impact of your innovation to diverse cultures as a way of testing its value is brilliant.

Some consumers of and contributors to the transHuman Code platform will digest the content with an investor's appetite. It's a good thing, as you know, because innovation often needs capital. In that light, what are the areas in technology right now that you would say are the most promising seeds of potential growth?

Pentland: I have to say that, like anybody, I see things from a particular perspective, but the part that's most interesting to me is the combination of data, AI, and blockchain together. The reason is that you can see the emergence of an ecology of personal trusted data—based on sensors and IoT, blockchain-type systems to be able to make these things robust and trustworthy, and then AI systems to provide real-time insights.

So, it's really a renaissance of awareness that I see coming, and the renaissance is based on the ability to be able to project the future, based on evidence, and to project that future in a far more reliable and viable way than we've ever been able to before.

So, in the large-scale historical picture . . . we developed ways of speaking and language, and then began to write things down so we could preserve knowledge, and then the story of the last century is really about being able to have broader communication, and as we now have sensors, measurements, data everywhere that is becoming much more trustworthy, much more something that you can build firmly upon, and AI to be able to help us manage that data, we begin to have a broader and more active awareness of the world.

It happens in trivial ways. So, for instance, it's now very infrequent that someone will become lost, that they won't know where they are, they won't know where they're going, that they won't know how long it will take to get there and what the best way to get there is. That used to be a huge problem for everybody all of the time, and, essentially, now we have these projection tools that use the data that we all use to be able to coordinate ourselves far better than ever before.

Or search engines. Finding information used to be a lifelong hassle. Where you find information for a plant that will live here, or the right food for this or that? So, now we can find out facts, you can find out availability, you can find out pricing and trends trivially, and people do this hundreds of times a day. So, it's just this increasing awareness of our environment that we've never had before, ever, that's occurring. I think that's beginning to change us, not make us less human, but giving us a broader awareness.

And you could look at a lot of the pessimism that people see as an offshoot of that awareness, and you become aware of a broader process in society, of events. You can see the bad along with the good. And, a lot of times, I think people in the more wealthy sections of society were insulated from that. You never saw people in serious or grave or harsh situations, and now it's everywhere, because the tools we have are so much broader and so much more powerful than they ever had before.

That's a really fascinating take on pessimism as an offspring of the awareness. It's articulated so well and it's accurate. We see that playing out. In light of this renaissance of awareness, what human traits now become that much more significant? What are the traits needed for us to translate the renaissance of awareness into productive innovation, into effective innovation that actually progresses areas that need it right now?

The biggest transformation is the way that innovation is changing. It used to be that you could look at a person's life and then come back 1,000 years later and the same people would have pretty much the same tools, the same life. That's not true even year on year now. So, what that means is that a more static way of thinking about things is important. It used to be that laws were meant to be permanent. And you would have a career in a particular area or a body of knowledge that's your expertise.

All of those become silly, because things change so quickly. In a certain sense, people have to let curiosity and entrepreneurship become their greatest guide, and that requires an inordinate amount of courage. It takes courage to transform yourself and adapt into these new elements of reality and to try them on for size. So, the comfort of doing the same thing that your father and mother did and doing the same thing you did a decade ago is evaporating, and people must pick the next higher energy. But we have to also know that it's a much more rewarding role than we've ever played before.

So, instead of being people camped in the plains, we have to be mountain climbers now. And for that, optimism is important. We know, as a matter of fact, that optimists are much more successful at fixing problems than pessimists. If you're going to have a successful society, one that adapts to great challenges, you need a society of optimists. They need to believe that something can be done and that together they can do it. If that's not the case, that's a society that's likely to disappear.

Authors: That's well said. The psychology of optimism is that it opens us up to an inordinate number of opportunities that we wouldn't have otherwise seen if we were pessimistic.

Talk to us about the Brooks Brothers initiative you've been a part of and then, secondly, the Berkeley music initiative, since we're on this topic of being optimistic and using an entrepreneurial mindset. These are two innovative endeavors using humanity and technology to progress in areas that are very encouraging.

Pentland: Here's a little context first. On a very abstract level, the story of the last couple centuries has been the emergence of the machine, and the organization as a machine. So, the factory, grades in schools and degrees, professions, all those things are static structures to support, essentially, an industrial society. But the industries are changing so rapidly that this system is beginning to evaporate and the world is becoming more fluid. Technology needs to support the fluidity and make it something that is comfortable and human.

The problem that has arisen is that with the great machine of organizations came the suppression of the individual cultural variations. Everybody had to be the same. So, the opportunity now is to move to a world for diverse small cultures to work together, to be able to challenge the machine. It's not quite the libertarian view of things, but it's a mixture of the very old, where you had small, intimate communities, and the very big where they can band together. Interestingly, that's exactly what technology like blockchain and the Internet and IoT allow you to do.

The reasons for the big corporations, at heart, was to lower transaction costs. You could make widgets cheaper and better if everybody did exactly the same thing. But now we have the computerized systems, these digital systems that allow us to coordinate with each other much more fluidly, in a much more trustworthy fashion, so that you can have lots of variations and it still works incredibly efficiently.

So, what we're doing is, A: building those systems like AI, blockchain, and IoT, but also, B: looking for the place where the transformation then goes. So, the canonical thing is lots of craftsmen, individual people who have special ways of doing things, being able to make a living by assembling their work for worldwide markets without a big machine and corporations in the middle.

Music is going that way. The big record labels, the big distribution channels are all evaporating. Currently, that means that the artists have no way to get paid. But what we're building at Berkeley is a digital rights platform where individual artists can contribute digital media, music, and get paid for it when it's used, and it's a very fluid sort of medium that doesn't require big, monolithic corporations to be successful. Not that corporations can't help, because certain things require concentration of resources, for instance, to pay for concerts and setting up concerts or for advertising or things like that, but it becomes a much more fluid thing where individual artists can communicate with the world and, in a much more successful way, be paid.

The same is true of Italian business. Lots and lots of small mom and pop artisans are increasingly being forced out of the marketplace because of coordination costs. They're unable to compete with the big firms because they're unable to service, individually, worldwide markets. But if they're able to coordinate, they could make world-class products, products that work better than the mass-produced ones, and get those delivered. So, we're working now with some Italian companies so that suits at Brooks Brothers will be sourced from artisans around Italy the same way we're talking about musicians being able to sell their music across the entire world in a much more fluid way.

And many of the people that sponsor our research here are interested in this new, much more fluid economy. Of course, we have people like the IBMs of the world, but also some of the large financial institutions, like UBS, who are interested in this much more fluid culture, this much more fluid future, where the big corporate promise of reliability becomes irrelevant. With a fluid economy, you have almost perfect reliability, and perfect flexibility as well.

Authors: Hearing you describe this new economic reality, it's like a hearkening back to the Renaissance in some ways.

Pentland: I think the most interesting thing is that traditional human activity—the act of creating—is becoming increasingly enabled by the technology as opposed to being suppressed by the technology.

Authors: I think that very point is counterintuitive to what people are expecting, that technologies like blockchain are going to take away from them as opposed to give back to them and or recognize what they're doing and doing well already.

Pentland: Yes. A third endeavor we're undertaking is the notion of digital banks that represent individuals. It's now actually possible to build little bits of software that are completely invisible, that capture all of your digital footprints and save them in a digital savings account for you, and by bounding people together through these cooperative banking structures, you can end up with people who control a copy of their data and number in the tens of millions, which means they have negotiating power.

And just like in the past, we had unions to be able to band the workers together to negotiate with companies, where we had financial institutions to band together to be able to become important actors through retirement funds and other sorts of funds, we can begin to see the same thing emerging in the datasphere, where people band together to become consumers, but more importantly, enablers of digital services through the ownership of copies of their data.

So, now what we're seeing is, I think, data unions, people banding together, and that's the hope, that this begins to balance some of the mega data owners, the Googles and Facebooks of the world, so that communities have much more control over their data. I think that's one of the really important issues that's out there, and I think the notion of forming these cooperatives, which are already legally enabled, and the technology is not difficult, is a very promising avenue for this sort of thing.

Authors: That's fascinating. Your insightfulness and positive outlook is certainly inspiring. But there must be something that worries you enough to keep you up at night?

Pentland: Well, I think the negativity that you mentioned earlier is certainly disturbing. When people feel threatened, when they feel broken into small, defensive identities, they're not very good at finding solutions. If everybody's trying to hold on to every last scrap and suspects the other guys, then the basis for successful collaboration evaporates, and that keeps me up at night. I think of it a little differently than other people perhaps. I don't think of it as good versus bad, which is what seems to be the dominant way people think about it. I think about it as society being frozen in these camps, which prevents coherent solutions to any of the global problems.

The truth is that most of these global problems are pretty simple to solve, from a practical point of view, except that you can't get people to agree. So, the real problem is not global warming, water shortage, food distribution imbalance, things like that; it's how do you provoke a successful, productive discussion? Because if we could do that, we could address important things pretty straightforwardly.

Authors: Sandy, you spoke about the importance of two institutions, the World Economic Forum and MIT, and the significance of civic engagements. What troubles us is that, in the absence of a governor of technology, there are very few controls in place, if any, about what is developed when and how it's applied and what the displacement effects could be of it. How do we best address this?

Pentland: Yes, I hear you. That's a good question. I can tell you what MIT is doing. I'm not sure how this translates to the WEF. But what MIT is doing is becoming much more integrated with the entrepreneurial environment. So, thinking about having the impact, starting companies, I run an entrepreneurship program that particularly focuses on start-ups in the developing world. The point is that real world conditions and science have to become more happily married, and people need to be aware of these broader problems.

So, it's still the case that most of MIT is just not aware of how most of the people in the world live or what are the real practical problems, and they have this fantasy, industrial silo fantasy, where you do something in isolation and you throw it over the barriers to the next silo. No, actually, people need to read everything from the theory to the practice and weight different elements of that differently, but it needs to be something that includes that entire range.

And I'll just say for myself, I've founded companies and put them out in the real world where they serve real people, and today I would like to see lots of people doing that. I don't see why our efforts need to be siloed. I think that's a perverse, unintended consequence of this industrial specialization urge. If we were all "jacks of all trades," then we would see that we can contribute to each of these important stages of change.

Authors: One of the things that we are really intrigued with is the incentives that you're offering to your MIT students, grants of $5,000 to $20,000, to support their good ideas. Can we assume that you are expecting of them that careful consideration of the ethical impact of their technologies is prioritized in their address of practical problems also?

Pentland: While I am not a fan of separating out ethics, I am a fan of innovating ethics, which means that if you have all the different communities interacting, you will care about ethics. And what does ethics usually mean? It usually means that some community is getting shafted. So, if you're developing science, and it sounds like a really bad idea to some community, you've got to listen to that. It has to be something where everybody agrees it's a neutral good idea. If you hear some people say, "Nuh-uh," then I think you have to listen to that and respect it, and that, to me, is actually the core of ethics.

Authors: Sandy, we have thoroughly enjoyed this discussion and are grateful for your continued support and insightful contribution to the transHuman code project. ▲

Conversation with Beth Porter

Cofounder and CEO of RIFF Learning, researcher and lecturer with the
MIT Media Lab and Boston University Questrom School of Business,
and an artificial intelligence pioneer

Authors: Beth, you have spent your career envisioning and developing computer-enabled and online teaching and learning experiences. Now you have created a leading-edge platform at Riff that helps people and organizations innovate better through AI.

With your vast experience at the center of knowledge development and sharing, I think a great place to begin this conversation today is with your perspective on the importance of that intersection of innovation and collaboration?

Beth: Yes. From my point of view, people come up with really interesting ideas all the time. You're in the shower, you're riding on a subway, you're in your car, and those individual ideas are all really important germs for things that may eventually become game changing in our lives.

But the only thing that separates a really neat idea from something that's innovation is the ability to help other people understand it. For them to reflect on it and give you that feedback. For them to help you refine it so that it really is something that's not just in your own head but something that you can share and relate to other people about.

The whole process of coming up with new thinking, sharing with other people, communicating about it using all those feedback loops that you engage in as humans are really important for the refinement of the idea into the thing that arrives in the marketplace where people look and say, "Oh. That's actually really interesting and innovative."

I remember reading an article about Jeff Bezos, that he's been thinking about space initiatives since he was ten. And it's not like he's just been sitting around thinking about it by himself. He's been thinking about it and talking about it with people and reflecting those ideas everywhere that he goes. Feedback tells him he's going in the wrong direction or the right direction. That constant mechanism of feedback, that is what Riff is actually all about. That's what we do. We give people the opportunity to do that with their ideas.

Authors: Tell us more about your approach to convince more people to embrace collaboration when they are also concerned about protecting their intellectual property?

Beth: It's really difficult sometimes because I think people misunderstand how intellectual property (IP) should be defined. Most people believe that any great idea they come up with is protectable. I don't know if it's because we live in a litigious culture or what it is that has trained people. But very little of what we do in technology is actually subject to IP. And yes, we do in fact have to convince people that the sharing of ideas has more benefit than it takes away. It's not trivial. We actually have a hard time sometimes. I'll give you an example.

I know people who work in the autonomous car industry. In that space you're really talking about an enormous number of secretive activities. Right? A team cracks the code for how to track the movement of animals on the road when you're driving in the country. The car knows how to tell a raccoon from a shrew from a deer or whatever. The minute that this problem gets cracked, everybody thinks, "Oh we've got to lock that down. That's ours now and we're going to securely protect it. Because that's going to be the thing that makes or breaks us as an autonomous vehicle company."

But the thing that people forget is that there's more value in the sharing of those principles than there is in holding on to those for dear life. Not necessarily the underlying code or algorithm but the principles that you use and determining whether they're going to be applicable in multiple conditions, with multiple vendors, with potential partners. When you do that, you're never going to be able to go through that innovator cycle, which examines whether you've really solved the problem that you're all aiming toward.

Now I can't solve the whole macro business problem of IP and trade secrets. But we can and are influencing the process of person to person idea sharing. That's where we work.

Authors: You're tackling a social misconception. The Napster story, the Facebook story, an antagonist running with someone else's original idea. We know those stories and they're famous now. So we have this knee-jerk reaction to protect ideas. We say, "It's mine. I'm the one that's going to make millions off this. I'm not going to let anyone else do that."

Beth: One of the things that I'm noticing as a trend is people are less reluctant than ever to use open source code solutions.

What we're finding is that more and more things are built on top of open source solutions, even the most restrictive of them. And what we see is that the value is in the combinations and collaboration, the uniqueness of the solutions, the human capital that is being delivered, the way in which things are combined and created in a particular space to solve a particular problem that is unique.

So, some things will continue to be subject to very stringent protective licenses that will never reveal the secret sauce behind a compression algorithm, an AI machinery algorithm, or anything like that. But because of the way that code is evolving, because of the way that people are learning how to code coming out of college, a lot of these protections are just not going to be important anymore. The code is not going to be important. It's just going to be the way that it's composed and delivered and the service and the value that's put on top of it.

I used to run product and engineering at edX. I ran the Open edX project, and what I came to discover is that by using totally open software, such as ADTL and Apache License, people have created an enormous value for customers. Really building very little of their own code. And that's very telling to me.

Authors: That is very telling. In terms of the innovations that need to take place, what skills are the most critical in this sort of human-technology marriage?

Beth: One, it's really critical that you learn how to find value in other people. I know that sounds a little bit weird, but one of the things we have often found in previous generations of innovators is that single people don't really innovate individually. Individuals who understand how to create real value in the world always work with other people, and they have a key set of partners that they rely on to help them refine their idea providing critical feedback. People who are deaf to that, who can't hear that feedback, or who aren't welcome to it, or who don't invite it, they're not going to be able to be innovators. So anybody who has an innovation mindset knows how to work with other people. They have to.

The second thing I would say is that innovators know how to collaborate with them too. It's not just that you sort of convene meetings and get feedback from everybody, some positive, some negative and then you all just walk away. Collaboration involves riffing—yes, that is the origin of our name Riff Learning. It comes from very iterative, fast-moving idea generation. You know, being able to put ideas on the table, throw them out, bring in new ones. You have to be able to rapidly share, refine, and discard ideas. And if you're not good at that, or if you need things to be perfectly encapsulated and perfectly presented, it's likely that you won't be a good innovator.

The third attribute is the ability to embrace people who are really different from yourself, to value them because they're different. Not insisting that everybody be the same type of personality that you are is important. If you're a creative type, don't just work with creative types. Embrace the fact that somebody works totally different from you. Engage with somebody who has completely antithetical ideas of yours. Pursue somebody who has a totally different mindset or structure through which they think about problems.

Authors: While cultural diversity certainly expands the likelihood for different ideas to come to the table, it does not necessarily mean you'll collaborate with those who have a different approach to solving a problem, a different approach to getting a job done.

Beth: I worked with these great women who run the company Impact Seat. It's a diversity consulting company but not one that comes in and says, "You're not hiring enough women or people of color." Instead, they say, "Look, it's all well and good. Your hiring practices definitely should embrace diversity. But they should also reflect diverse thinking." So they're trying to help people understand that diversity comes in lots of different shapes and sizes.

Authors: Along the theme of collaboration, what can we do in collaboration with AI?

Beth: I teach a class over at Boston University in the business school, and the most revisited topic of the class is, "Oh automation. AI. I'm scared. I'm not going to have a job." And so we spend a lot of time demystifying the misunderstandings about AI and why automation is not something people should generally be afraid of.

One of the things we talk about a lot is changing the vocabulary and talking about augmentation. Augmentation suggests that you will be complemented versus replaced.

Many of the adult students in this class are in sales and marketing roles. Roles threatened by AI. And I tell them, "One of the original promises of the robo-advisor was that AI would become good enough to relieve you of the basic analysis so you can focus on a higher level set of activities and tasks that you'd never even get to because you were doing all this low-level stuff."

So when we talk about AI and we talk about the roles that people will transition to in their careers, the opportunity begins to feel more positive than negative.

Authors: How will AI help us thrive the most, where we haven't been able to before?

Beth: Frankly, I think what we've done in AI, up until now, is actually pretty trivial. Many of the targets being addressed are social engineering problems and some convenience problems. And that's a safe place to start. But AI is soon going to be credited for some incredible advances in fields such as medicine and genetics.

Authors: One of the underlying threads in this transHuman Code book is an acknowledgment that the human body is ultimately the greatest technology on the planet. Whether we're evolved or created, it doesn't really matter for the sake of this argument. We're merely making a statement about the human body right now, which is that it is the most technologically advanced system on the planet. We agree that AI can help us better understand and use this greatest of all technologies, our human body.

Beth: That's right. And that shouldn't be just for elite athletes who have had access to these feedback mechanisms for many years. It should be for everybody. The ability to self-monitor and understand the impact of changes in eating and exercising is so valuable and becoming available for all.

Authors: With your interest in the application of AI to our basic needs, what most inspires or encourages you?

Beth: An industry that I believe will benefit greatly from AI is food security. I think we should apply ourselves vigorously to innovation to solve food impoverishment in the United States and around the world. We have one of the wealthiest countries in the world and we still can't feed everybody. I believe that AI can play a key role in our address of this challenge across the world.

Authors: What can we be doing to demystify innovation so that more people will be encouraged to embrace it, as opposed to push back on it and/or run away from innovations, such as AI, as it relates to food and other industries?

Beth: I think it's just about education. One of the key things I work on as part of my role at Riff is communicating the importance of learning differently, breaking the belief that "Learning is learning, working is working, and nary the twain shall meet."

I'm continuously trying to help people understand just how symbiotic these two activities are and should be. If we allow people to more seamlessly blend their goals as learners with their goals as professionals, we'd all be in a much better place. I actually started my career as a mathematics teacher in high school and college. I believe in the power of education, but I don't necessarily believe in the preeminence or dominance that higher education has had in being the sole place where credentials can be issued and people can learn. If we had less structure there, I think people would be able to feel like they could learn, understand, and apply. At the risk of making an analogy to laundry, "It's wash, rinse, repeat. Over, and over, and over again."

I'm sure that you have had this experience: You go offsite to a workshop and you learn some things. You get up the next morning and you go to work having forgotten ⅞ of what you just learned in that conference. Literally. These mechanisms, these manufactured mechanisms for learning, as the only ways that we can legitimately learn, are not working. We need a much more rapid, much more entwined way of learning and working, and working and learning.

You never can presume who you're going to learn most impactfully from. Where it may be a professor you have in college, the more likely thing is it's going to be a mentor that you have in your professional life, and interestingly we don't really value or credit those people as much as we should. It's just not part of how we think about our learning.

I remember seeing a corporate slogan that said something like: "Lifelong learning. Learning for life." Or something like that. And I thought to myself, "Yes. If it were actually true." I think that it's always been a promise and never, ever been delivered on. Now we have the tools to deliver. We just need to be open to learning in new ways, and in new contexts. The data is available to teach us. We just need to learn how to apply it well and regularly.

Authors: Beth, knowledge is only valuable if it can effectively be shared. Thank you for your generous contribution to the transHuman code project and to the innovative learning that lies ahead of us all. ▲

Conversation with Leena Nair

Chief Human Resources Officer of Unilever, global champion for the establishment of human-centric leadership initiatives, including diversity and inclusion, for over 160,000 employees in 190 countries

Authors: At the core of the transHuman code is our objective to create a conversation among innovators—those who are conceiving, implementing, and using technology to enable dynamic change across our life ecosystem. The goal of this initiative is to provide a forum to advance a prosperous and positive relationship between humanity and technology.

In the range of roles that you held at Unilever, the advancement of the organization to establish human-centric leadership initiatives including diversity and inclusion has been a consistent focus. Today in your role as chief human resources officer, you're charged with the well-being, the contribution, and the future of more than 160,000 employees in 190 countries. I'd like to begin by learning more about how you envision your responsibility and the planning for the future of Unilever and for all of its stakeholders?

Leena: I've been in the HR function for 25 years. I believe that the next 10 years are going to be the most exciting for any HR professional. Knowledge can be matched. Skills can be matched. Small brands are beating big brands. Anybody with an idea and Internet can create a business. What differentiates business today is people: their ideas, their passion, their creativity.

I believe that the next 10 years will be so critical for the human resources function because we've got to see ourselves as leading the business and laying the road for the business, not following the business and filling the cracks that are left behind. In that regard, I see my role as central to helping this business navigate a fast-changing, unprecedented time for the next 20 years. This is brought on by the Fourth Industrial Revolution, influx of new technology. These unprecedented changes are also because people are living so much longer.

And, the fact that we are increasingly in a borderless world, where you've got to stop thinking about ownership of people. You've got to think about access to people. I see my role as strategically shaping the business to get ready for this unprecedented change. I see my role as helping the business see that technology can be used to define and complement the human in all of us.

The need to develop our human capital has not diminished at all. In fact, there needs to be more. I believe it can be achieved. Infinite human capital can be tapped into, if you use the power of purpose, the power of lifelong learning, and the ability to harness everyone's potential.

I describe it in three simple words, and that's my HR strategy on a page: capacity, capability, culture. If you do that, people with purpose will thrive in an unprecedented world.

Capacity is all about learning to work in new and different ways. Gone are the days of organograms, boxes; that's not a way it works. The world in squad teams, sprint teams, flowing people to where the problem is. Quickly regrouping to solve something together, which needs complex problem-solving skills. I call that capacity. Capacity is about— if you close down factories—putting a plan to rehabilitate 95 percent, 100 percent of your workforce. Re-skilling 100 percent of your workforce to do other things.

Capability is about continuous learning, lifelong learning, giving people the courage to learn. Culture is creating an environment where people feel they are the heart of everything we do. That we care about them. Because, like I was explaining on the panel, the whole world's talking about unprecedented change. Technology is coming. Robots are coming; 75 million jobs are gone.

Now people are scared, they're anxious. So, you've got to invest in their well-being, in their purpose as much as you invest in skill building and skill infrastructure. Just because you tell them every morning, "You've got to learn. You've got to learn, this thing's happening," they're not going to be motivated to learn. I've got to work with you to help you understand your purpose. What excites you? What gets you out of bed? How do you see that purpose coming to life in this world of unprecedented change?

I've got to care about your physical and mental well-being. Because if you walk through the door with depression, anxiety, worry, or fear of failure, then you're not going to be productive, happy, or ready to learn. So capacity, capability, culture, working on all three will help people develop with time.

For example, we're putting 100 percent of our workforce through purpose workshops. And people say, "Oh, why does somebody who's packing Lifebuoy soap, or somebody who's packing shampoo, why does he or she need to know about purpose?" They do. They do, because they need to see how their work is meaningful, and how it makes a bigger difference to Unilever.

So, 100 percent of our workforce is going through purpose workshops, and as I speak, more than 40,000 have already finished. They spend an entire day or two looking at what gets them out of bed. What's their sweet spot? What are their strengths? What is their motivation? What shaped their childhood? What were their principal moments?

Literally, they end the workshop with a full understanding of their strengths and development areas, and how what they passionately care about fits into what Unilever is trying to do. These are amazing sessions. We're having increasing evidence that people who've gone through this experience, their engagement is going up, as reported through the regular poll surveys and people surveys that we do.

We are putting all our employees through well-being programs. We say not a single employee should be more than one click, one call, one chat away from feeling well. This means employee assistance and support programs, and full measure. It means everyone has a chance for a physical examination every year and a mental health assessment every year, or every other year.

We believe in our well-being framework, which is physical well-being, mental well-being, emotional well-being. We make sure people get time with their families and do things they're passionate about. And purposeful well-being, which is what I said about people with purpose thriving. And last, but not the least, financial well-being.

So, we put every person to creating a well-being plan for themselves. So they, again, go through a couple of days in workshop, which we call the Thrive Workshop. Where they define for themselves, what is my well-being plan? One person might not get enough sleep. For somebody else, it's that they really need to lose weight or look after their nutrition. For somebody else, it's finding a buddy for exercise, what have you. We encourage them to create well-being plans.

They create these well-being plans, and then these are reviewed at critical times with their leader, who asks, "Hey, how are you making progress on your well-being plan? How are you making progress on your professional skill development?" We invest in all this as a foundation to give people the confidence to build capability.

Another thing we're doing is we've invested in a company called Degreed, the Silicon Valley company start-up that is focused on lifelong learning. It is a platform that curates all internal and external material available on topics, and allows people to access millions and millions of pieces of content.

So for example, in this workshop, individuals develop a thrive plan, an individual development plan where you say, "These are my strengths. This is my purpose. This is my potential. This is what I want to do. These are the three actions I need to do to develop myself." And all these are then available on the system, and Degreed has the capacity to analyze your strengths, to understand your role.

Using all AI and algorithms, it then figures out that, for example, Leena needs to learn design thinking. By looking at my strengths, looking at my role, looking at what my purpose is, it figures out that I need to be doing more in design thinking. So, every morning and once a week, I get a feed from Degreed, just like you get a Facebook feed or you get an Instagram feed that says, "Hey, Leena, these are the five things we think you should be learning, and here are some others who rated these things."

The program nudges me to go in and see a video or read an article, or just little bites. Because that's the other thing: people learn in snacks nowadays. Snackable learning is another term we use to describe bite-sized learning and the personalized feed that comes every day. But, I don't think we can simply create infrastructure and give people a lot of learning to do if we haven't invested time in finding out and understanding what their purpose is. And helping them figure out the development areas, or helping them feel the courage and confidence to learn.

We can have all these fancy learning programs that you read about every day, but they are not gonna work if we've not worked on your motivation to learn. I wish it were as simple as saying, "Courses are available. Go and do it." You need to be motivated to learn. There is a mindset of saying, "I'm ready for lifelong learning." We invest, therefore, in creating the foundation.

These are some of the things we're doing to harness everybody's potential. Capacity, teaching people to work in new and different ways, and to tap into the wider ecosystem of people. Thinking about access to skills, not ownership of skills. It's about capability, both functional capability and leadership capability. And it's about the culture. Creating an environment where people feel they're included, like they care about us.

I'm so glad you're writing this book, because I get really annoyed when leaders say people are our most important asset. I say, "Show me the proof." I have to see two things. I have to see investment into people, and I have to see time that you're setting aside for people.

If CEOs are not spending between 30 percent to 50 percent of their time on people-related issues, their teams, people who work with them, the broader workforce, employee engagement, employee training, employee leadership development . . . If our leaders are not spending time on these things, don't tell me people are your most important asset because they're not. You're neither putting in the time or money.

I am also concerned about the imbalance of investment being made by corporations with greater contributions being made to new technologies than the re-skilling of the workforce.

Authors: In the developments and the implementation of your well-being program, are you learning more about what concerns your employees have about how their relationship is going to be impacted by technology? I know that you're using AI in a positive manner with Degreed, and I look forward to learning more about that.

Leena: When doing Degreed our recruitment is entirely digitized end-to-end. Our people service stuff is entirely digitized. If you come and see my room, you'll see two big screens on which, in real time, I get analysis of the feeds of our employees, what's on their mind internally—our internal social media—and what's on their mind externally on all the external and social media. I'm having a look at it and having fun with it all the time. I just look at it once a day to see what's keeping people in my app.

Now, when we do the analysis about mental health and well-being, the calls are all anonymous, but we look at totality. What were the kinds of concerns and anxieties people had? We realized financial well-being was on a lot of people's minds. That's why we improved our well-being framework to also include financial well-being. Not just mental, physical, purposeful, and emotional, but also financial.

So it gives us a very good indication of what's concerning our people. It supplements the work that's coming through our UniVoice, which is our Unilever voice of employees. So, you're absolutely right, the work on well-being connects with the work on what's coming from our people's survey, what's bothering people. It connects the teams.

When the automation and facts are all back, significantly, we implement a specific change program. So, for example, in some of our factories, we have a digital factory initiative going on where they are using more and more predictive algorithms to decide when to do the particular chemical, how to pack. It's going on in three big factories, and we put in place automation training. We've run learning webinars for the new engineering organization. We appointed a technical skills leader and backed them up.

I know I'm running ahead of myself, but in early March, my CEO and I are signing with our biggest union, a commitment to re-skill them.

Every time we make a job redundant, we re-skill them. And my personal vision is that 100 percent of my workforce will be rehabilitated, re-skilled, re-deployed. I know my team says I'm crazy. "They're never going to get to 100 percent." But I want to keep that as an aspiration. If we come in at 95 percent I'll be okay, but I don't want to relax. I want to try for 100 percent.

Because which employee am I going to look at and say, "I don't care about your re-skilling, redeployment, or reengagement with the workforce?" So we really want to lead the way among businesses and open sourcing our learning content on technology automation and make it relevant, and freely available to people everywhere. We also want to use technology and big data, as it unlocks that ability to do this at speed and scale.

So, for example, portability of learning records, credentials, and the facility for people to share these across the recruitment platforms are essential. Using blockchain to provide a way to validate credentials and skillsets. We're exploring all of that. So my vision is everyone in Unilever, and hopefully by inspiration the rest of other companies, will be relevant for the future.

Our jobs may get redundant, but our people will be relevant and right. When we re-skill them to the jobs that are going to come when we re-deploy them in the right ways. We reengage them in different forms and capacity to work with Unilever or with other partners. So that's my vision, 100 percent of our people are going to be relevant, and a lot of jobs are going to be redundant. Does it all make sense?

Authors: It makes perfect sense. Not only is the aspiration admirable, but the implementation of strategy to achieve it, we can hear it clearly in what you're doing. It's not enough to make the statement; it's the action that truly matters the most. How are you undertaking that responsibility to predict what the roles of the future are going to be? How are you applying science to that process?

Leena: Fantastic question. We work closely with the World Economic Forum. We go country by country, and we have a partnership with LinkedIn and we have partnerships with other retail and FMCG companies, which Walmart and I are jointly co-chairing. Because 19 percent of the world's employment sits in the consumer goods sector. So, a Unilever, PNG, a Walmart can't look away; 19 percent of the world's employees work for us.

Either these people work directly for us or they work on the stuff we're asking them to do, which is to go to shop. So, go and visit the trade, or go feed the industry, then deliver to people. We're going country by country, using AI and the help of LinkedIn, to predict the map for that country. Let me give an example.

We know that truck drivers are increasing; are their jobs going to stay or not? But you may not be able to move a truck driver from being a truck driver to being a data scientist tomorrow morning. It's not a bridging role. So we think about each job, and we say, what are the adjacencies? What can this person do, what skills are sensible and don't need them to transition to be something that they simply cannot?

So for every profession, every skill, let's say packing soap—I'm making this up—we look at the five things that the packer could actually do, seeing what he could or she could bridge into, and then we focus on those skills. That's where the worker's purpose becomes important, because the person says, "You know what? I don't want to do that. I want to do this. I don't want to do stuff with data, but I absolutely enjoy doing something with our brand," or whatever it is.

So we use the work we're doing on purpose and development to create these maps. So country by country, we're mapping our people, mapping the professions, and looking for adjacencies. Looking for where we can bridge, because you can't bridge from A point to every point. You have to bridge from A to B.

We haven't done this in all countries across the world, but we've picked six or seven of our countries where we are piloting and experimenting this approach with the unions on our sites. So, we've had long chats with all our trade unions, and they are working with us to make this happen. Then when I succeed in those five or six countries, we'll replicate its scale across all 190 countries.

Authors: Thank you, Leena, for sharing your passion and wisdom with us.

Leena: I am extremely passionate about this subject of human advancement, and I appreciate the opportunity to have my voice heard. As human beings, we have the ingenuity, the talent, the intelligence, and knowledge to create opportunity from all the disruption and change we are experiencing. The transHuman code is a valuable vehicle for us to be able to contribute to and be guided by. ▲

Conversation with Dr. Enrico Fucile

Chief of the Data Representative, Metadata & Monitoring Division, World Meteorological Organization (WMO), leading the gathering and predictive analysis of crucial climate data from the 191 member countries and territories

Authors: Enrico, the global climate has become a unifying subject among citizens across the world, and never before has there been so much focus and attention on meteorology and environmental science, which for you has been a life's work. The attention and the responsibility that's placed on you and the team has never been greater. Prior to joining the WMO, you were with the ECMWF (European Centre for Medium-Range Weather Forecasts), which of course is the world's leading center for atmospheric predictions. You were also a member of the Italian Meteorological Service. Let's begin with how your role has been evolving with the increasing demands for insight into meteorology, environmental science, and the strategies that we should be adopting?

Enrico: Yes, I have had the opportunity to work at both the national and international level. My role has always been in providing the technological support to operational and research activity, at weather environmental agencies, in general.

My work has changed significantly, moving from the direct support of the national agency to the international organization like the ECMWF where the organization was serving the needs of 34 member states. Again, it was, for me, another big jump to move to the World Meteorological Organization.

The WMO is a UN specialized agency with 191 member countries and territories. When I was working in the Italian Service, I was directly serving the needs of civil protection, aviation, agriculture, and others. At ECMWF I was providing data and service and coordination to the member states.

Now, in the WMO, I'm more providing governance, guidance, and coordination on a global scale. This is a cascade process, where the WMO is providing the set of regulations and framework and coordinated work on regional and national organizations.

Continuously, without any significant interruption for years, we have been collecting this data. So it's very exciting to be part of this worldwide structure, and it is a big technical and technological challenge.

When you work for a national service, you are aware that there is a dependency of your national service on data that is regulated by the WMO and provided by other countries and other organizations. But you are also aware that other countries benefit from your data, so we are members of a big community.

Now in all these years, technology has changed a lot, as you know. Most of the technology of the 90s is now obsolete. Systems have been changed to adapt to dramatic increases in volume and how we use the data. We have seen also that not only is there an increase in the volume of data, but a sudden increase in the demand for weather information. The demand is driven by the more frequent occurrence of catastrophic events, connected with climate change, and also by another important element, which is the evolution of the users.

Users are more technology aware. They demand more. We have better technology, are creating new opportunities, and striving to feed the more ambitious demands from the user. Again, more advanced offers and more ambitious demands, so it's a cycle where the private companies are dynamic and are pushing this cycle forward.

The increase in the offer of environmental information is not always good and positive, because it tends to come with great responsibility for the technology to reassure the user that the information is good, that we are providing authoritative information. The WMO strives to bring an authoritative voice to weather information.

What I've seen in the evolution of my role in these years is that in the past the demands for the big production of data were not as great. Today, there is a demand to make this information publicly available, so the policy is a bit different. There is more of an emphasis on the use of web technologies and on the provision of services. We spend a lot of time discussing how to improve the discoverability of our authoritative services and data in a web that is flooded by so many kinds of meteorological data of unknown providence.

Recently, we had this discussion with member states, and it was requested that we conduct a review, a report on emerging data issues. The conclusion was that data is a means to an end, not an end in itself. Its true value lies in how effectively it is used to meet societal needs. In the short term, it supports the response to severe weather events, over the middle term it facilitates the planning and preparedness for weather resilience, and over the longer term for historical climate insight and assets of impact.

So, in short, we are flooded by data, and the priority is to understand the value of the data.

Authors: You've been assembling data since the beginning of record keeping, from across the world. Is the most valuable role today how you're interpreting that data, and how you're modeling with that data the probability of events occurring?

Yes, the most valuable role of WMO is not only in keeping data and analyzing data, it's also in providing standardization and guidance and coordination between the member states and their activities in a way that they can work together to assemble all the weather information.

Let me tell you a story about the most successful program of WMO. I don't know if you know the World Weather Watch, WWW, but that's the name. It was defined the first time by WMO in 1963. This is a program that had the aim to build a global infrastructure for the observation of the atmosphere in real time. At that time, the need was seen because of the opportunity presented through technological change. We had the opportunity to gather data from space, from the first satellite. This was a great opportunity for weather, because it was the first time that you were able to see the atmosphere from space. WWW is still operational now, since 1963.

WWW is considered an example of international cooperation. It doesn't have any equal today. We are talking of 192 states working all together to exchange information, to make independent and collective observations. The WMO's success of meteorology infrastructural globalism is unequaled.

We made a big change in the beginning of the 2000, where there was a decision to build a better system to expose not only weather information, but also other environmental information. This is working now. The name is WIS, WMO Information System. And now the data landscape is changing dramatically. Now we have big data, cloud computing, artificial intelligence, and new data gen technologies, so now we are at the point of having another big jump. In fact, we are now designing our new system, the WIS 2.0.

The motivation to move to a new system is the increasing demand of authoritative weather, water, and climate information. This information is to be made available to public, private, and academic users. This is connected with the UN Global Agenda, including things like Framework for Disaster Risk Reduction and the Paris Agreement, all global priorities. So today, we have the need to make more authoritative information privately and publicly available, and discoverable. This is not a trivial task, because the information we have available from our member states is very complex to ensure that data is of the required quality.

Authors: This talks to our third question, which is what technologies are you using most effectively, and do you feel that you have access to the resources you require to be able to interpret and to engage together with all stakeholders on a plan of action? Is there anything that you feel you're missing at this stage?

Enrico: Right, the problem when we talk about implementing technology on a worldwide basis is that our implementation is usually very slow. We employ a technique where we try to implement technology in some countries, then we export the technology to other countries, and then create the regulation where all the countries can come together and work together.

We think we have enough technology to achieve what we want to achieve at the moment, but there are problems, mainly with establishing trust, establishing the information as authoritative.

The way this works is by making regulations and allowing our members to develop the applications. So for example, we are going to plan to work with new technologies for exchange of data. The technologies will also include, perhaps surprisingly, social applications such as WhatsApp. And we also strive to make standards in a way that all the members can develop their own applications. This is important to our way of work.

Today, we believe we have enough technology to provide the services that we need to provide.

Authors: We are particularly curious about how you engage with the governments of the member countries, as well as with corporations, academic institutions, and even financial leaders. Recognizing that you have the information, you have guidance and direction, how then do you transmit, engage, communicate, and ultimately achieve your desired effect? What is that process?

Enrico: Well the process is well regulated, because we have permanent representatives of all the countries in our organization. They are agents of their government, so in that sense we have a direct connection with all the governments. As an example, in June, we are going to have our congress. In the congress, all the representatives for the countries will come to discuss our strategies and align on our implementation. This is our WIS 2.0.

Authors: Is that system, that process, satisfactory to you, or can it be enhanced?

Enrico: Our view is changing. Our members are asking us to change in a direction where we have a more holistic view and where we are more oriented to deliver services to the users; everything is changing. We are trying to make a more effective and agile organization.

Authors: That is encouraging. We do believe that there's a more heightened level of awareness among corporate leaders and finance leaders today about the impacts of our climate and the effects that our actions are having on the environment, which affects us not only personally but also professionally. We will be interested to witness the evolution of that process. We've got to believe, as well, that the new digital technologies for communication are certainly enhancing your ability to be able to engage.

Enrico: Yes. You know digital technologies are the foundation of our work in WMO. We are always trying to improve and to go on with new technology. As an example, we think of the farmer, in the less developed country, with very few resources to install weather stations, a data center, unable to predict future weather events. You know, to make weather predictions you have to have a powerful data center. In some countries, you don't have this capacity. Today, technology in communication is able to cover these gaps. We are able to deliver also in remote places of the world; the information is vital for the citizen that is living in a very high risk area for natural disasters.

As I said, we are well aware of the changes in the landscape, but one of the big changes that is happening now, and it's going to be discussed during this year, is partnership with private companies. Today the discussion of private/public partnership is very strong at WMO because of the technology. Satellites today can be made very small. Cheap to build, cheap to launch, and cheap to keep in orbit. So, private companies today, with private capital, can run space meteorological programs in full with very little money. This is a significant advancement!

There are several private companies that can now provide data with a special business model to be commercialized for weather forecasting. We have developed a model based on free exchange of data. And our member states are freely exchanging data between them. So this is a new element in the global weather enterprise.

Authors: What will the effect of that be? Will it be complementary? Will it be competitive?

Enrico: Yes, there are policy challenges, but there are also, in a sense, technical problems. I can tell you, because you could think to build a marketplace where weather information is produced not only by government.

In the past, satellite programs were run by public, by the nations, not by the private sector. We are working to understand how this market is going to be built.

There is also a lot of discussion about blockchain. New technologies will be at the base of this market for certain. I really believe that this is going to be one of the most interesting evolutions of technology. I don't think we are quite ready yet. There is really the need of some written policy to try to make a quality global weather enterprise. What is positive is that the will is there to build a positive relationship between private and public in the global weather enterprise at different levels.

We are now organizing workshops and, during the next year, a big conference involving all the stakeholders, to try to understand where we can go with the technology and also with policy. Policy and technology will come together to build a different global weather enterprise for the greater benefit of all.

Authors: Enrico, this was a most enlightening conversation. You've provided us with a great deal of insight that we didn't otherwise have. And we're sure that readers of *The transHuman Code* will benefit greatly from your contribution.

Enrico: Thank you. I am confident that we too will benefit from the contributions of your transHuman code community. ▲

Conversation with Marc Firestone

Marc Firestone, President of External Affairs of PMI, global advocate for innovative change, and co-leader of one of the most dynamic corporate transformations in history

Authors: Marc, for over 30 years you've been working with Fortune 500 multinational companies, and you've been witness to significant transformative change through adoption and implementation of new technologies. I'm interested to begin this conversation talking about you, and we'd love for you to characterize your experience with the evolution of technology and how it's been guiding your professional activities.

Marc: In terms of my experience and activities, I think the most positive and desirable experience I've had with the evolution of technology has been access to information via the Internet and the developments that have digitized enormous libraries of materials. Literally taking physical libraries and digitizing them, as well as putting them on the Internet or making available via the Internet information that originated in the digital format has been hugely significant to me.

I've found it enhancing to me professionally, given that a lot of what I do entails research, studying things, understanding things, as well as personally. Billions of people have found it enormously useful that Google, and other companies are making the world's information available to everybody in a useful format. It's just been huge.

I remember in the 1970s when I first saw earliest versions of what has now become LexisNexis thinking just how extraordinary that was. And I remember showing that to friends in the '80s and '90s when they started to put, I guess what became the Nexis side of it, in newspapers and other media allowing us to be able to look up a review of a movie or of a restaurant, for example.

And of course now we have Google and other search engines. That to me is the most powerful influence of technology for me professionally.

Authors: Well certainly the definition of search has evolved dramatically. We had to commit much more consciously to what we were going to look for and how we were gonna look for it. And the time…

Marc: Well, the cost of information, you probably have the data, but the cost of information acquisition is a tiny fraction of what it used to be. I will say, at the same time I still enjoy physical libraries and find an enormous amount of information in my *Encyclopædia Britannica* at home. So I think there's certainly a place from my broader social perspective for both books on the bookshelf and countless masses of data available via the Internet.

Authors: The technology to which you're referring is both a simultaneous affording and demanding transformation of traditional corporations, and PMI is not alone in evolving its products. Certainly service companies across all industry sectors and geographic regions are evolving. We'd be interested to learn about PMI's philosophy and history of technology adoption.

Marc: Yes. I suppose that there're two aspects of it. One of course is the one that is driving our complete transformation right now, which is technology that enables us to provide products to men and women—who will otherwise keep smoking cigarettes—that are a much better alternative for them for their health and for their satisfaction.

And the other aspect is the use of technology in essentially running the business. With regard to the former, I would say well over 20 years, and most intensively and extensively for the last 10-plus years, we have looked and found ways to provide satisfaction for people who would otherwise keep smoking by heating tobacco or by aerosolizing liquid that contains nicotine.

It's been surprisingly difficult to do that. To generate an aerosol from a solid substrate of tobacco is very difficult if you're not going to light and combust tobacco, which is of course how a cigarette works. That is a relatively simple process of igniting the tobacco leaves. To create an aerosol that is at a sufficiently low temperature so as to not generate the array of harmful chemicals in the quantities present in cigarette smoke while still reaching a temperature that'll produce an aerosol released in nicotine is difficult.

It's a matter of chemistry and physics. So that's been a huge part of our research and develop efforts. And along with that has been the enormous focus on the biology, the toxicology, and the clinical investigations to look into a new product, to check, double check, and quadruple check that it is in fact a better alternative. So through all of those aspects, I would say that the philosophy is to make sure that we are being extremely precise and rigorous in use of technology. On the other aspect, we are also focused on how we use technology . . . increasing the efficiency of internal processes, of interactions with business partners, the whole sheer of B2B relationships we and other multinationals have. Of course, these processes are quite complex and have enormous amounts of data associated with them.

And then I would say spanning both aspects of the technology in a product for consumers, and in the technology running the business, is of course technology that enables us to focus on consumer information technology and communications technology. Then we can be absolutely contemporary and ideally even forward looking from a technology standpoint in terms of meeting and anticipating consumer needs and demands.

Authors: Well, it appears that you are using available technologies to their fullest potential. We did want to talk about the evolution of digital technologies for the communication industry and how the awareness level of issues, challenges, and opportunities has been heightened greatly.

We're curious about the communication that's happening both forward to marketplace, to your customers and distribution channels, as well as that information that is coming from the users and from other channels of distribution back to you. How has that evolved? How has your relationship with the customer evolved through the use of new technology platforms?

Marc: This is an area that has been changing, in my view. It's an area that has changed, I would say, a lot in the last three to five years if I think back over my exposure to the company. We've been in the cigarette business, which of course is an old category from the 19th century. And our company has its roots in the 19th century and has grown dramatically I would say since the 1950s, 1960s largely internationally through geographic expansion, some acquisitions to enable geographic expansion, but largely through organic growth.

We have had strong B2C relations, and the consumers have certainly known us through brands and the branding, but there has not been the type of interaction with the consumers that some other consumer product companies might have. In the last three to five years, while we have been introducing new products that are part of what we call our "smoke-free future"—these are products that don't burn, don't have combustion, to put it differently—we have been much more focused on how to interact with the consumer. And a lot of those interactions are and should continue to be one-to-one with somebody who is in our sales group who understands the new products, who's able to meet with the men and women who smoke cigarettes and understand their interest and explain the new product.

And around those core human-to-human interactions, we are focused on communication tools that current and developing technology enables, whether it's the mobile phone, cell phones, computers, or otherwise. So we are certainly entering into all these new areas of communications.

Now, by virtue of selling the products that contain tobacco or nicotine, we often encounter restrictions on our use of certain media tools. But against that baseline, certainly we're very interested in being able to hear from consumers who have questions, who need more information on how to use new products. Again cigarettes are simple to use and don't require people to learn how to use them.

With a new product that is essentially relying on electronics for the functionality there are inevitably all sorts of questions: how do I clean the new product, how long does the battery work, what do these flashing lights mean, et cetera. And so, just as other companies that are selling products that have electronic components do, we want to be as advanced as we can in the level of communication.

Another aspect of the communications industry of such that is hugely important to us is miniaturization. When we have our products that heat tobacco, they do so through, I would say, sophisticated electronics, and the ability to use those electronics to heat tobacco depends on miniaturization. The communications industry has been extraordinary in the last 30-plus years in miniaturization, and that is something we track. And secondly, these products involve batteries. We're not selling large products that you plug into a wall to use, we're selling products that are portable, that are convenient, that are easy to use where your electronics are typically 100 U.S. dollars, or 100 euros or less.

And so the development of ever more powerful, long-lasting batteries exist of course in many industries, but I personally think of the communications industry as driving much of the innovation in batteries, and that's hugely important to us as well.

So the communications industry has made us aware of consumer functionality in many respects, both in terms of the active communication of exchanging information, as well as the technology to enable that act of exchanging information.

Authors: Marc, PMI has been heralded, not only for the vision but for the actions that you've taken to date toward a smoke-free future. What stage are you at in that process right now without asking you to divulge trade secrets?

Marc: Well it's enormous, it's frankly something that maybe five years ago I would not have foreseen, and I say that having spent many years in and around this company and our former parent company. By that I mean five years ago, 100 percent of our revenues, if I can recall correctly, were from the sale of combustible tobacco products. And then in 2016, beginning 2017, we stated unequivocally that our vision is to replace cigarettes with alternative products that are a better choice than continuing to smoke. It was not a strategy of diversifying into other industries, and thereby to have cigarettes a smaller part, by percentage, of our total revenue or to merge with another company and sell off cigarettes. It was to do this action that we think is the most dramatic we could have figured, which is to say we have at Phillip Morris International roughly 150 million men and women around the world who are smoking our products.

There are about one billion people in the world who are smoking cigarettes, and we want to see the day come when there will no longer be combustible tobacco, no longer be cigarette smoking. In terms of some examples in the end of 2017, about 13 percent of our revenues came from these new products. So from 2015 through the end of 2017, we went from essentially 0 and 100 percent, respectively, to 14 percent and 87 percent, respectively.

Today, we have entered over 40 countries with our new product where we have roughly six million adults who were smoking and have converted to our smoke-free products.

We have said by 2025, we would like to see 40 million of our consumers be consumers of the smoke-free products. So we're at the stage of being fully behind this new strategy. It is the predominate strategy to characterize how focused all 83,000 of us in the company are on this strategy. We are certainly not at the end of it. We still are very much building the new business, but we are certainly well beyond the hypothetical phase, well beyond that to the phase where it's truly what the company is doing.

You would agree that to test a company's sincerity, the best thing to do is to look at how they allocate their resources. And I can say that over the last three-plus years, we have moved large resources from our traditional core cigarette business to our new smoke-free product business. Our sales force, sales and marketing activities, and of course research and development activities are where we are putting the company's resources to ensure that what we're doing is a business success, and therefore it is a success for our consumers, and therefore it's a success for society at large.

It's, I think, an unusual situation of alignment of a business strategy, which is to say we are replacing our product of circa 2014 with a product circa 2015 and beyond by replacing cigarettes and smoking products. We are doing that so people who will otherwise continue smoking have a much better choice for their health, and third, that as more and more people around the world switch from a path of continued cigarette smoking to using a noncombustible product, it will have benefits for public health, and therefore society at large.

For me it's very exciting and enormously complicated, which makes it also fascinating. And because of the health aspects, it's also enormously satisfying to be part of the venture.

Authors: Marc, it's an incredibly complex undertaking for any organization to transform itself as you are. What was appealing to us when we were first introduced to the company's vision and strategy was it really embodied the core principle of our transHuman code Initiative with placing people first in the relationship and utilizing technology to its fullest potential.

You're engaging regularly with global thought leaders and influencers. How is the rest of the world responding to this? What's the range of responses that you're receiving?

Marc: The range of responses, that's a good way to ask it. I would say the range of responses is from support, enthusiasm, interest, fascination, so very positive. I would say sometimes the response that we encounter is, "I'm sorry, tell me that again? I don't understand. Aren't you in the cigarette business? How is this possible?" so a very reasonable, quizzical response.

And I emphasize very reasonable because even for me having been at the table literally as we worked on this, at least once a week, I have to pause and remind myself of the full magnitude of what we're doing to some element of skepticism to cynicism. Skepticism of "I can't believe you're really doing this" to cynicism, "you're a cigarette company and there's got to be a trick here."

So I would say it goes across that spectrum. My experience, though, is it's much heavily weighted certainly to the left of skepticism, if it's a spectrum that's immediately clear and it's immediately applaudable, to the other end of the spectrum of, wait, there's gotta be a trick here. So I would say that it certainly leans heavily to the left of the middle on there because people understand that fundamentally we're a consumer products company, so we're in the cigarette business traditionally, and of course that's among the most, if not the most, controversial of all consumer products.

But as a consumer products company, I think the vast majority of people understand the proposition that if you're making a consumer product that has an imbalance of stated neutral reduce that, you should work to reduce that imbalance. In other words, if you're making a widget that people have been using, that people might enjoy for whatever reason but the widget has negatives, of course if you can remove or reduce significantly those negatives, you should. And society should not only permit that, it should enable that. And that's fundamentally what we're doing in neutral terms and in our specific terms is to say, "People are smoking cigarettes. In any given year of a population of cigarette smokers in a given country, maybe 5 percent to 10 percent will quit, which means 95 percent or 90 percent will continue smoking."

So take the 90 percent, that means in a slice of time, statistical look, 9 out of 10 people currently smoking in country A will keep smoking. Not necessarily 9 out of 10 individuals, but of the population, 9 out of 10 in that population will continue smoking. And cigarette smoking causes disease, a function of quantity and duration, which means as people keep smoking, their risk of serious disease increases.

So we're saying for those 9 out of 10 people, we're gonna give them an alternative choice, a much better choice of products that do not have cigarette smoke but are sufficiently satisfying, although not risk free and still addictive and are dramatically less likely to cause disease. And for all the controversial aspects of the tobacco business, the cigarette business, we think that that's just a fundamental proposition that people who would otherwise keep smoking should have access to information about less harmful alternatives.

When we meet people, that's the key point we offer. We acknowledge of course that we are in a very controversial sector and that there's a lot of controversy with our company. What we're asking for, though, across the spectrum of reactions, even among those who are cynical, is a fair assessment of the data, a scientific assessment of the scientific data, ultimately in the interest of the men and women who will keep smoking. And yes, it is a business proposition for us, but it also is a business proposition whose outcome is directly related to the health and well-being of the billion people who smoke and of course all those around them.

Authors: Marc, when we began to study the future of health and looking through the lens of how technology can help us improve health, we can honestly tell you that we did not expect that we were going to be discussing this subject with PMI. So too were we surprised to learn that there were a billion people smoking around the world.

So we were pleasantly surprised that not only was there a smoke-free vision, but you are affecting it. And through this transformation, you are committing financial and human resources, and drawing upon a broad range of external resources to create this revolution. And it truly is revolutionary. We know that others will question, as we did, how PMI could factor in to a conversation about the future of health, but you may in fact, as an organization, have the single most dramatic impact on the health of our population going forward, and we are so appreciative to be able to learn what we have through you and to be able to share that through the transHuman Code Initiative.

Marc: Well thank you. Thank you for saying that. I appreciate it so much. The company, my colleagues, have embedded this mission, the vision of strategy, into all that we do. We're very excited about its success. ▲

Conversation with Jaeyoung Lee

President of KGMLab, Professor of Civilization and Technology,
former Member of the National Assembly of the Republic of Korea, TV
commentator, and technology start-up advisor and investor

Authors: Jaeyoung, during your term in public office, South Korea was first awarded premier ranking as the most innovative country in the world, which is admirable in and of itself, but that achievement has since been repeated four more times through to 2018. From your experience, both in the private sector as well as academia, and of course as a member of the government that received this first recognition award, what do you attribute this accomplishment to, and what does it tell us about South Korea's future and the technological revolution?

Jaeyoung: This recognition was attributed in large part to Samsung, who has received more U.S. patents than any other companies other than IBM. How this investment into innovation trickles down the supply chain throughout our country is very interesting. And yes, while South Korea is advancing, especially in technology innovation, more work needs to be done where this sort of innovation culture can be spread to other industries and the mind of all the country's population.

Ninety-five percent of Korean people now have smartphones, which is the highest penetration in the world. Korean companies, as well as government, push very hard on 5G technologies. We already have three major telecommunications companies investing heavily on 5G. Korean people are very interested in using technology, and their appetite for high-speed Internet is the highest in the world, I would say. I think we have a very good infrastructure, both in human capital as well as in hardware, to move forward with new innovations. In Korea, we have 50 million people, and the size of the country is just the right size. It's small enough and big enough for a lot of these technologies to be tested. A lot of Western technology companies actually come to Korea to test their new innovations. I think we are well provisioned in global markets to adapt and test these new technologies moving forward.

But at the same time, I also think that in order for us to move forward, we need education reform. We still require our students to enter top universities on their ability to take college entrance exams well. It does not really reflect their innovative or creative mind yet, but the system has been in place since the beginning of the Republic of Korea, which is about 70 years old. Reform will take a huge effort, and it'll produce a huge social clash in the future. I think that will be the biggest challenge to advancement that we face right now.

Authors: Interesting. Well, one of the other questions we have pertains to technologies for which South Korea very early adopted, and perhaps this was driven by Samsung's transformative technology and infrastructure, and that is artificial intelligence. A recent McKinsey study reported that almost half of paid work as we understand it today can be automated. Now, obviously there's a period of training, scaling, and implementation, but we have the capability to do that now. Yet, still fewer than 50 percent of CIOs globally across multinational companies were installing or even testing AI. Do you think that South Korean companies are more advanced in the application of transformative technologies like AI?

Jaeyoung: Well, I think our economy, perhaps like any other advanced nations or advancing nations, is split into two groups of companies. Those who do invest in future technologies, and the ones who don't, or who can't. Samsung is obviously one of the top examples of companies who are investing heavily in these new technologies, and they are establishing AI institutes throughout the world.

Other companies like LG, SK, the big names, they do invest heavily, and they do have deep pockets to do so. But also at the same time, the Korean economy is still heavily dependent on labor-intensive work. These companies are not doing well in terms of investing in new technologies. I mean, they have not even invested in basic automation, so that poses a problem at this point.

But I think, given that our economy depends heavily on these heavy, big conglomerates who are able to invest in these new technologies, moving forward, I think these will eventually trickle down to the other industries and will make a huge impact in the way we work and how the companies produce their goods.

Authors: Is there an open dialogue today, do you believe, either within the government or within other bodies that are watching the evolution of that relationship between people and machines? Is it part of the conversation?

Jaeyoung: Yes. This government, which came into place in the summer of 2017, created a committee on the Fourth Industrial Revolution. Within that committee, they have discussed a lot about the future technologies, on how they can be applied to the current industries in everyday lives. Within that discussion, they talked about how some of the regulations are hindering these technological advancements.

There already is an active interest as well as work being done by the government at the highest level, because this committee was created by our country's leadership. The ministries are paying attention, and there are many forums in South Korea, private forums, small and big, especially in academia, discussing these things.

Authors: While Samsung certainly was a significant influencer in the selection of South Korea for the innovation award, there have been other credits and recognitions made of the country and its technological prowess. That really permeates all generations and across all business, social, and commercial activities, doesn't it?

Jaeyoung: Yes. For example, when Apple came out with the iPhone, Korea was not even one of the countries to receive those phones. I think the first model that was introduced in Korea was iPhone 2, so after a year and a half in the U.S. of smartphones, in 2009. But that was also the first year for Samsung to come up with smartphones. So, in just 10 years thus far we have the highest penetration rate for mobile phones in the world. It shows again the very high appetite that we have for these new technologies.

Authors: Yes, and we think that we can be looking to South Korea and what is evident in its sensitive relationship between humanity and technology. It is estimated that there will be 75 million jobs displaced and as many as 135 million new roles created by 2022. It's that 60-million-person gap in between that's most disconcerting. We certainly have unprecedented challenges for how 21st-century employees are prepared. For companies whose primary business is technology, it's much easier for them, of course. There'll be less re-skilling and more training education for application of deeper skills.

Jaeyoung: The larger companies apply technology that is relevant to their own industries, to improve their industries moving forward. I think there's a natural selection of what kind of technology becomes popular within certain countries based on the industries they already have.

At the same time, I think while we are well prepared, or we are open-minded to these new technologies, we do still see huge potential social clash. Recently, one of the leading technology companies in Korea tried to introduce carpool service apps, which was met with huge opposition from taxi drivers. Thousands of taxi drivers just poured onto the street, with some taking the extreme action of setting themselves on fire.

The company ultimately scrapped the idea of introducing this service indefinitely. It just shows how a technology or application of new technology can threaten people. During that incident, the government created a special committee to discuss how to find a solution to appease both parties. This is just one example illustrating the difficulty for society to move forward into the Fourth Industrial Revolution.

More than half of our industries are small- to medium-sized companies who are very labor intensive. Some of these companies will innovate. But what can we do with the workers in companies that don't?

It's a huge challenge. When I ran for a seat within my electoral district, I would estimate 50 percent of the population in my constituency are not part of the companies who are willing to hold their employees in place and guide them, support them through these changes. As a politician who has to be elected to affect change, I need to bring forward solutions that will offer benefits to my entire constituency.

As long as politicians in democratic society need to appease the general population, I am not sure how the new ways of doing things, or preparing for the future, can be done in the best way. People are thinking about it, and we cannot deny this huge change that must happen, and it will happen whether we are ready or not, but this political system that we are a part of is a huge roadblock to move forward.

Authors: One of the ways to address the challenge is by more effective collaboration between governments, universities, and business. Co-op education brings corporations and educational institutions together in a collaboration that benefits all: the employers, the students, and certainly the schools. And the government can support that too. What do you see as the future of educational institutions and corporations collaborating in South Korea?

Jaeyoung: Well, I think it's going to move forward, and I think the co-op model that started in the United States and Canada is definitely being spread throughout the world. Korea's leading universities are definitely doing that with corporations, but I think the corporations need to be more willing to share with universities what type of workers they need, and be transparent about it, given that it is in all of our interest to do so.

But I think they also need to take another step, almost a leap, which is to think about what type of workers they need 10 or 20 years from now. What I see when I teach at the university right now is that these universities are willing to work with corporations by producing students or educating students for corporations and what skill or work skill they need today.

But, I am not sure whether these work skills will be available, or even be good work skills 10, 20 years from now. I think corporations are not being successful, or they're not really thinking about what that picture's going to look like. University, while they might be thinking about it academically, they're not applying that to their current education system. So I think that needs to be done. That really needs to have a long-term vision, and corporations need to invest more with the universities in future work skills, which might not pay out right now or tomorrow, but will definitely help the entire society to prepare for the future.

Jaeyoung: We're already seeing that in medical fields, for example, but we need to spread out across the other basic industries, I believe.

Authors: We think that there are potentially two areas where the universities can play a role in their relationship with corporations. One is, as you've shared, the importance of guiding the educational institutions on what's expected to be needed, and expanding on the cooperative programs. Second, potentially is playing a role in helping to re-skill the workforce, so bringing the power of educational experience and expertise to the organization. Do you see that as something being possible and appropriate in South Korea?

Jaeyoung: Yes, absolutely. I think what I would call a post-postsecondary education is something that is very much needed. So these workers need to be continuously, consistently reintroduced to do things in new ways. I think university kids definitely play a role in that, but I also think, for example, Korea is heavily invested in private education. Now those private education institutes are more geared toward educating secondary students in preparation for the colleges. But they have the ecosystem where they can bring in students to be taught.

I think a benefit of private institutes is that they can be more nimble than universities, because universities have so many regulations they have to live by, and it's very hard for them to change curriculum. It takes a long time for them to make any decisions.

The private sector educational institutions can play an even bigger role if they can work together with corporations. I think corporations need to look at them as one of the ways to re-skill their employees and future employees.

Authors: Well, we concur with you and think this is an area that you in your next electoral term can certainly have an impact on. And we think it's going to be important that the governments be proactive and encourage that. We are most concerned about the small- and medium-sized business. The large corporations have the resources, both human and financial, to dedicate to the study and reassessment. Unilever is an extraordinary example of how the individuals can be placed first in the hierarchy of change. But we worry most about those companies that are investing every dollar and every human resource hour into just continuing to do business in the present, as opposed to being able to prepare for the future.

Jaeyoung: I believe that government needs to learn how to change their perspectives. They really need to look at things from a very different angle or a variety of angles, right? Our government is so heavy on clamping down on them, because they see it as something that is disturbing the public education. But if they can change their mindset and say, "Since we do have this industry, and since we have this huge well-established eco-system, maybe we can use it for good," right? If they see other opportunities and other chances to give them the role to play for building a better future, and have these private education sectors to work with small- to medium-sized companies, like you said, to re-skill themselves with government funding or whatnot support, I think that could create a good movement, or at least to ease the pain in between.

Authors: You were a member of the World Economic Forum (WEF) Asia team, and through your engagement with the WEF Global Leadership Fellow Program, you've been witness to the tremendous evolution of the technology shift to dominate the main stage for the forum. What do you believe the role is of multilateral institutions such as the WEF in the management of the technology-humanity relationship?

Jaeyoung: When I joined the WEF in 2009, they had a whole team dedicated to bring the technology companies to the global stage. I believe Google, when they were not Google as we know now, was first introduced in the World Economic Forum, as well as Facebook and other technology companies.

I believe that what these major multilateral institutions can do today is to really dive into the social issues. I think these social clashes, evidenced in the car hire sector here, for example, will happen not only in Korea, but this will happen everywhere in the world.

The government needs to play a major role to mitigate these problems. I think those can definitely be examples other countries can adapt or use.

Those issues and those solutions or discussions need to take a much larger portion of the WEF Forum, for example. I think these multilateral institutions definitely have a role to share these experiences globally, and I think that's where the future lies.

Authors: Jaeyoung, at the root of the transHuman code, as you know, is the core belief that we have to place people and their positive development at the forefront of this relationship between humanity and technology. What are your closing thoughts on how we can maintain the most human-centric approach to technology and the developments for the people of your country in the future?

Jaeyoung: You know, being a politician, I think the government needs to play the number-one role in this area. Not with frivolously imposed regulations discouraging innovations, but we definitely need to realize that these technologies, if they are being developed and introduced and applied to society without any sort of mechanism that is monitoring, developing, and even using or auditing these technologies, can be dangerous.

One of the examples is the development of AI—that really needs to be closely monitored. Again, not to discourage innovation, but to make sure that it is being applied and used in the right area for the right reason. I think the government, with the help of civil society and also the private sector, needs to be more aware of what they can bring to this discussion. Our future really hinges on it.

Authors: Jaeyoung, we are so appreciative that you've shared your time to contribute to the dialogue. We hope it's just the beginning of our conversation and our collaboration on this initiative. ▲

Conversation with Mohit Joshi

President of Infosys, a global financial technology industry pioneer, leading innovative banking, financial services and insurance, healthcare and life science transformation around the world

Authors: Mohit, you've literally worked all over the world at the intersection of financial services and technology, and in a wide variety of roles with Infosys, for almost 20 years. During what's been the most dynamic period of business transformation that we've ever experienced. We thought it would be interesting to begin our conversation with some of the insight into the transformation that you've personally experienced working at the forefront of this dramatic corporate change.

Mohit: If I go back to the start of my career with Infosys, I'm looking at the time when the original dot-com bubble was bursting. I think there was a little bit of a pessimism at that time. It appeared like digital and ecommerce was going to change the world, but maybe it was a false promise.

Actually, if you look at the past five years—the amount of transformation we've had, the new businesses that have been built, the brand-new industries that have been created—it's truly remarkable. From reasonably humble beginnings and a deeper interest in technology, the commercialization of the world wide web, in the past five years we've seen things grow at a much more rapid pace. If you look at companies across the world today, I'm struck by the fact that almost everybody is trying to do three things at the same time.

First, people realized that they need to be competitive in a global economic context. People still have to focus on their cost-to-income ratios, they still have to focus on the unit cost of their production, but companies across the world are focused on industrialization. Many large companies have not been automated front-to-back. They have a number of manual processes resulting in a number of inefficiencies, which have resulted in bloated costs for many of these producers.

Banks, insurance companies, manufacturing companies, and retailers all need to industrialize. Which really means adopting automation and AI in a fairly significant way for core operations, and looking at utilities where possible rather than trying to do everything themselves. Focus on using the public cloud, if possible, as a source of cost advantage as compared to building out your own data centers.

The second phase is digital, and I admit that digital obviously is a very broad term. Let's unpack it a little bit and begin with a focus on the customer. There has been a huge focus on customer experience in the past five years, and it's not just for external customers but even for internal customers. How do you get an Apple-like user interface? How do we get an Apple-like experience? This has become very important for most companies. Apart from the experience piece, the focus is also on data. Many large businesses today, much more so than 10 years ago, are completely into data. If you look at the likes of a Google or a Facebook, but increasingly even if you look at financial institutions, you look at retailers, these are businesses built on a deep understanding of data and of how customers will react in certain situations.

I think data is the second piece of this digital revolution that we've seen, and then there are other components. There are components around innovation, there are components around cybersecurity. Again, even on the digital side of things there's a huge focus on the cloud. But while on the industrialization side, the focus on the cloud is for cost efficiencies, the focus here on the digital side is to provide superior experiences—cloud for speed.

Really to become more industrialized and to become more creative, all companies also dramatically need to change the ways in which they work. We spoke a little bit about this. I think the agile and DevOps revolution is changing the world, and this has a major impact on how companies are organized, how they deliver on their core businesses and their core outputs. This also means that there has to be a huge focus on the education or retraining of employees.

Obviously, there are technology components to what I've seen in the past five years. We can be talking about AI and automation and blockchain and the mid lever experience technologies, but if you step back a little, everything that has happened in the past five years is focused on significant cost efficiencies and industrialization. It's focused on a significant embraceability. It's focused on transforming the ways in which we work.

Authors: Financial services has always been considered a core pillar for the advancement of all market economies. Your work has seen you introducing and providing tools to both developed and developing markets, and we're curious, from your experience, how are the developing markets capitalizing most effectively on the application of digital technology? Specifically, how are they impacting financial inclusiveness with their populations?

Mohit: I believe this will obviously have a huge potential to bring financial services to a much broader set of the population in developed countries, where up to 60 percent of the population is still not covered by the traditional banks and they're still not covered by the traditional insurers. That's a significant opportunity, because it obviously reduces the cost of distribution dramatically. Branches, and in fact an entire network of agents for instance, can be established in rural areas, simply using the mobile phone. The mobile phone has been remarkable in allowing small businesses and individuals to participate in an organized financial system.

If you look at it from an India perspective, having this uniform identity or biometric for the lab has been significant. Certainly one of the biggest challenges in opening up a bank account in India, for instance, used to be the fact that you couldn't identify yourself or you couldn't give a fixed Google address. Certainly with the check identity, you've got 1.2 billion people who can provide sufficient KYC to the bank, and therefore open and operate a bank account.

Over time, the data accrues from all the financial transactions that these people started in the system. With this, they have a credit history and can obtain loans much more easily. If you tried to do this using the traditional branch network, it would be almost impossible. If you had to rely on the government to provide a traditional identity, in terms of a driver's license for instance or a passport, it would be difficult because in most of these countries less than 10 percent of the population would have access.

The fact that you can now prove your identity in a digital method, and that the efficiencies have reduced the cost of servicing an account to a few pennies per year from a few dollars per year, has had dramatic impact. That's meant that a lot of people can participate in this.

Authors: What have you witnessed as a difference in how governments, both in developing and developed countries, are addressing and facilitating financial inclusiveness?

Mohit: Governments are focused on financial inclusion. I think it makes a lot of sense. For one, it increases the tax base side because there are more participants, especially in developing countries, but even in Western Europe and the U.S. The more you pull the economy out of an informal sector into a formal sector, you expand your tax base, you expand your potential pool of individuals who can participate in the economy. I feel that the West has been balanced by a concern about government having access to your data, which is why you haven't seen a modernization of the Social Security number system or you haven't seen a similar biometric program coming about in Western Europe.

I think over time, and after building the right safeguards, this could become extremely important. As of now, some financial inclusion in the West is focused on people living at or below the poverty line, and the efficiency and the effectiveness of programs like food stamps to help this segment of the population. I feel that increasingly, the focus will divert to a mix of inclusion and education, because that then covers a much larger segment of the population. As many as 40 percent to 60 percent of people in the U.S. would be unable to come up with $400 at short notice. I think this statistic highlights the importance of both the inclusion and the education aspects of what needs to be done in financial services.

Authors: Yes, we agree with you. AI is bringing dramatic change to our work roles. You have over 225,000 employees in more than 50 countries. How is your company envisioning and preparing the team members for what their roles are going to be in the future?

Mohit: We are optimistic about the future of our business because I haven't met a single company, bank, or government that intends to spend less over the next 10 years in technology than they spent in the previous 10. I think we can take it as a given that the spend in technology by companies, by governments, and by individuals is going to increase. This obviously gives us a significant opportunity, because we're building many of the frameworks, foundations, and systems that are used by some of the largest companies in the world. From a company's perspective, we are likely to see a secular growth in business spend in technology.

We do have an imperative to ensure our people are ready with the newer technologies, but also in the newer paradigms for working. For instance, earlier you might have done work in a traditional waterfall method of software development, but now you move to agile. We may be developing systems and applications and supporting them in a client's dedicated data center, having to do it in the cloud. This really means that we have to be able to retrain our people, which means that we have to give them the tools, platforms, and physical locations so they can learn, and we are doing that.

We also have a fairly young workforce; the average age in the company is only 27. Therefore, there is a lot of optimism and a lot of bottom-up push. It's not like we have to push our employees to learn; our employees are actually pushing us to make newer tools, technologies, and training paradigms available to them so they can learn. For example, we decided to start a program training people on the software to develop autonomous driving vehicles. We tried to open it up to 100 people, but we had 7,000 people show up who wanted to participate in that program. The hunger to learn is high in the company, so that, along with the fact that we will see increased spend in technology, makes us quite optimistic about our future.

As far as the broader, systemwide implications are concerned, I think we have to be optimistic. Even though automation and AI will result in job losses in certain pockets, on the whole I think AI and automation will be a huge force for powering job creation. We have seen that already with the gig economy, we've seen that with brand-new roles. For instance, who knew there could be a social media consultant 10 years ago, right? We are seeing a dramatic increase in living standards in huge parts of the world. Billions of people have been lifted out of poverty over the past few years, again largely by the implementation of technology and globalization.

I feel that new jobs will be created. It's not like people are running out of things to do or people are running out of things that they want. I feel that while it is obviously hard to quantify at this phase, the jobs may come out of sectors that don't exist now. The arc of technology over the past thousand years has all been developed in additional goods and services being available, and a higher standard of life over the long run. I don't think that this wave of AI and automation and digital will be any different.

Authors: Infosys has a network of research labs and innovation hubs partnered with leading universities and educational institutions all over the world. One of the concerns that we hear voiced frequently is the ability of secondary and postsecondary institutions to be able to keep pace with the innovations that are driving the new skills required in this workplace.

How can you envision technology companies collaborating most effectively with schools, beginning with primary through post secondary, so the students can be best prepared for the present and for what's coming?

Mohit: Yeah, I think this is a very real challenge. The fact is that there is a huge demand for new skills, and the demands will only grow over the next few years. We want to make sure that we don't create a digital divide, which I think is to your point as well. Obviously, training in computers or advanced programs and data science are available in universities across the world, but we want to make sure they're available to everybody. Over here we've been working with some organizations like Go.org, for instance, which are focused on computer science education in the U.S. We've been focused on working with similar nonprofit organizations in India, such that at the grade-school level and at the secondary-school level we are working to provide the teachers and the materials needed to build the workforce of tomorrow.

Then, at the university level, here we do have formal relationships across the world because we hire from them. It's important for us to build those relationships in many countries. We bring the faculty of those universities to our campuses, tell them our perspective on the way we see the supplemental jobs shaping up over the next few years, and give our feedback on how their programs and training material need to adapt to the needs of tomorrow from business and from governments, such that they can start incorporating these inputs into their curriculum.

Lastly, what we're also doing is creating a significant number of platforms ourselves that a number of our clients are using. We then go into the retraining of the existing workforce. We have the materials available. We've worked with some of the largest banks in the world, for instance, as they've trained their employees in big data technologies. If you look at it across every segment of the demographic, whether it is school-going children, young adults in colleges and universities, or the existing workforce, we have created a program and means of engagement such that we can spread this gospel of technology to the widest possible audience.

Authors: Within your organization, we would anticipate that not only do you have greater visibility into what re-skilling is going to be required for your clients but also your own organization. What is the process of informing and then engaging your current workforce on what their roles could be going forward? Beyond the projects that they're working on at present.

Mohit: Sure. I think what we have to do is make a call. The way we look at it, you divide technology into three horizons. Horizon one is technology that is in widespread use today. These are tools like Java, for instance, or the reporting technologies that we have where the bulk of our people are currently engaged. Then you look at stoppages—horizon two technologies. These are more modern digital technologies being used now. You have big data applications, big data platforms, the work being done in front-end design, and the work being done in the automation space, for instance. These are the hard technologies of today.

Then, as we look forward into the future, we go to horizon three, which are the technologies coming up behind us that will be relevant in the future. For instance, advanced manufacturing, augmented and virtual reality, and advanced AI. These are the technologies that will shape the future, but are in the nascent stages now. Then, we look at the distribution of our work. How much of our work is being done in horizon one? How much is in horizon two? How much is in horizon three? What is the workforce that we have catering to each of these three waves of technology, and how will the distribution of the workforce between these three waves change over the next one, two, three, five, 10 years?

For instance, if today we have 50 people working on horizon one, 40 people working on horizon two, and maybe 10 people working on horizon three, five years down the line, horizon two will become horizon one, and horizon three will become horizon two. Therefore, you need to have a clear picture of how the new world works. Partly it'd be by retraining, partly it'd be by hiring new people and training them, and partly it'll be the lateral hiring experts that we do.

We have a fairly clear perspective one, two, five, and 10 years down the line how the demand for technology is going to look, and based on that demand, what are the adjustments we need to make in terms of re-skilling and refactoring and hiring? Obviously, we monitor this plan on a regular basis, and we modify it as the environment changes. There's a fairly sophisticated way that we have of trying to make sense of the world, or trying to make sense of the technology changes, and calibrating our own recruitment and training to be able to respond to that effectively.

Authors: We think it's fascinating to understand what your process is. We're curious, of the businesses that Infosys consults and provides service to all over the world, what percentage of those clients do you think understand the scope of change that is fast descending upon them today?

Mohit: Look, quite honestly I think most of the large companies realize the pace of change. I think they realize that they're dealing with a new set of competitors, they're dealing with a significantly empowered customer, and they're dealing with high expectations from their workforce as well. Especially as the nature of the workforce starts to change with more and more Millennials entering the workforce. I think people realize, especially at senior level and the board level, that the world is changing.

I think the speed of change, however, is still taking people by surprise. I think changes happen faster, certainly in some industries, particularly in retail. There has been a wave of creative destruction that has happened in the industry, and some of the largest players in hospitality, in retail, in automotive didn't even exist 10 years ago. In some other industries, like in resources, insurance, even banking, the landscape has changed.

People do realize the pace of change and the change they need to make, but the pace of change is not uniform across industries. I feel like the pace of change is only going to accelerate, especially with the location of data, with the newly empowered customer, and the new tools that we have because of AI and the public cloud. The pace of change in the next five years is going to be much more dramatic than the pace of change even over the past 10 years.

Authors: What about so many of the smaller companies, where you have the key principals literally in the trenches? They're not afforded necessarily the opportunity or financial resources for envisioning. What guidance can we give to those businesses that perhaps don't have capacity, don't even know that they would benefit from receiving guidance, direction, and services from companies like yours? How can we inspire and influence them to reach out, to learn the most about how they can be innovating?

Mohit: I think there's both a positive and a negative. The positive is that there never really has been a time like today when you have information literally at your fingertips, and you have a clear sense of what your peers across the world are doing in terms of the adoption of technology. Really it's all the resources available at our disposal. There's never been a time like this, and there is no excuse to be ignorant. There's a lot that people can pick up in terms of the cutting-edge trends.

I also feel that, in a sense, technology has become a lot cheaper. Not long ago, setting up your own data center or buying your own hardware or building your own applications was something that only the largest companies in the world could do. Today, start-ups use the most cutting-edge technology and deploy it globally for pennies on the dollar. The opportunities for small businesses also are significant, because suddenly you can compete with much larger players. In reality, for small companies, one huge advantage that they have is that there isn't much of a vested interest dedicated to the straightest goal. There is, therefore, a significant opportunity for small companies to transform themselves much more easily without the organizational inertia large companies have.

On the other hand, for large companies, I think the challenge is how do we adopt innovation such that we can stay nimble? A lot of times, as you know, the key barrier to change in companies doesn't come from the fact that the technology is too complex or that it is hard to adopt. It is that people are resistant to change, and in small companies it is sometimes easier to push through the change. I feel it is a time when small companies, new companies have some significant advantages in terms of how quickly they can grow. We have no doubt that the smartest of them will get the investors to become the billion-dollar corporations of tomorrow.

Mohit, your guidance in and of itself is inspiring. Earlier this year you were selected as the Young Global Leader of the World Economic Forum, which is a great distinction. Congratulations.

You've had an opportunity as a YGL over the last four years to become more engaged and involved with the WEF. We're curious about your thoughts regarding how organizations along with the World Economic Forum, like the UN, the WTO, and other multidisciplinary institutions, can play a role in fostering a higher level of prosperity and equality with technology. What has been your experience through the WEF and other organizations in that regard?

Mohit: I think the role of WEF is very important because the WEF is truly unique in terms of its ability to promote multi stakeholder dialogue. Typically, if you look at the UN for instance, it is still largely an inter-governmental body. If you look at many of the trade associations, like the Confederation of British Industry in India or the Chambers of Commerce in the U.S., these are very much institutions where the primary membership is corporates.

The WEF in that sense is unique because it has an almost equal representation from governments, from corporates, from nonprofits, from start-up enterprises. The quality of the dialogue is richer because it allows multiple people to take part. From my perspective, being part of the WEF, being a YGL has also given me access to a network that is truly multi-stakeholder. Particularly, I do a lot of work in the CBI as well. I feel like many of the institutions are important, especially in this time of change, in allowing multiple shapes to express themselves in providing a significant amount of research material and knowledge to be used by participants, and could be available as a sounding board in a time where there is a lot of change in a lot of companies, and individuals are looking for advice on patterns and trends and ways of working.

Authors: Yes, and is that something that you're consciously thoughtful of as each year progresses—how you're going to be able to draw upon, learn from, and contribute to the groups that you belong to?

Mohit: Absolutely. I feel that it's a significant opportunity to meet people, to understand different points of view, to be able to develop one's own point of view, and to be able to give back. To be able to give back to the community from which I have learned and gained so much.

Authors: Well, this is another core pillar of our transHuman code initiative, and as we shared with you, the vision for the transHuman code actually began as a conversation between Carlos Ferrara, my coauthor, myself, and Tim Burners Lee three years ago at Davos. What was discussed and envisioned then was a series of articles around the importance of communicating to all who were or could be using technology, about both the challenges and the opportunities with platforms that existed, and those that could. The resulting effect was the development of this conversation that we've been having for the last 18 months with a number of individuals. Now those conversations will come to life in the form of a book, a broadcast series, and open online dialogue in June. At the core of this is the desire to keep the human at the center of gravity in every relationship, especially the ever-evolving relationship between humanity and technology.

We have a desire to affect the pursuit of equality through the application of technology, which we're being witness to in increasing volumes.

I'm curious, as you continue in a leadership role at Infosys, how does this premise factor into your actions? How are you being considerate of how humanity can best benefit from technology?

Mohit: I feel that technology provides us with a set of tools, techniques, and platforms that are very powerful. At the end of the day, humans have to play a central role in technology. We have to play a key role in shaping the technology, rather than letting the technology shape us. The past trends have been positive. With everything we've managed to accomplish in the past 15 or 20 years—the huge increase in productivity, the billions of people being lifted from poverty, the explosion in global growth, and new industries coming about that we didn't know about—I feel we have to be optimistic about the future.

So far, the arc of new technologies like artificial intelligence leans toward empowerment. It leans toward giving us as humans the tools necessary to do new things, to detect and analyze patterns in data, for instance, and it makes possible revolutionary insights that could make our lives better. My perspective is an optimistic, human-centered view, quite like yours.

Authors: Well, we're not at all surprised, but encouraged, to hear that. Mohit, this was a rich interview for us. We're appreciative. What's interesting is that you've enjoyed 18 years with Infosys, which is a lifetime for millennials when they contemplate how long they're going to spend with a company. You've been witness to such tremendously dramatic technological innovation during that time. You've been looking forward for a long time, as a result of the organization and the role that Infosys plays in this technological revolution. It's going to be interesting to continue to have this conversation with you as we progress.

Mohit: Thank you. I look forward to that conversation because it is very important. It is important to all of us as we try and make sense of a changing world. ▲

Conversation with Daphne Kis

CEO of WorldQuant University, a global, online, entirely-free university dedicated to advancing education; a technology investor and pioneer; as well as a global champion for women entrepreneurs

Authors: Daphne, in your career, you've been a CEO, publisher, producer, investor, and ultimately you've been a nurturer of innovation during the formative stages of our digital revolution. And your history suggests that you've always been an agent of change. We'd like to begin our conversation today by talking about the change that you've witnessed during your own personal transformation.

Daphne: Thank you. I appreciate the opportunity to speak about this critical dialogue and the point of view you're bringing to it. I couldn't agree more on how important it is that technology makes contributions to society, including in areas such as education where we are focused. I have spent over 30 years in the tech world that, in its early days, was focused on enterprise, business-facing software.

This era was followed by considerable consumer-facing innovation, of which we all have unarguably been both the beneficiaries and the victims. Importantly, platforms are now globally conceived, implementation is not burdened by the last mile access, and the size of the so-called pipeline has expanded exponentially. Essentially, the promise of technology is being fulfilled more than ever, and it has transformed the world in which we all live.

But human endeavor is still behind it all; the drive to make everything more and more accessible is simply human nature. At a new level, we are seeing the fruits of decades of work and it's transformative. As the saying goes, the greatest change is the rate of change, and we've all lived through that.

Authors: And how does the pace of the first 30 years compare to today? You truly were a pioneer with PC Forum, a leading executive tech conference, which provided the platform for so many collaborations that have occurred throughout the short history of this digital revolution. How are you personally managing the speed of change?

Daphne: Well, the only way to manage change is to build upon it. Everything is constantly evolving and agility has to be built into products we see and use today. In the past, enterprise software was built to be stable, but stationary. Building business-to-business software in the cloud gives us the ability to regularly iterate on products and to update the software. Built-in feedback loops are now integral to design and to implementation. That's something that we value very much in the work we're doing at World-Quant University.

As technology takes center stage, the ability to iterate and be responsive to that speed of change is all you can hope for. To actually enable the life-long learning everyone is talking about, human agility has to be nurtured. This is a significant paradigm shift, in terms of how people think about the delivery of education. Responsiveness has to be in the DNA of the learning experience, in a way that's quite different from the past.

Authors: Well that would be yet another way that you are differentiating WorldQuant University from other educational institutions. We really just skimmed the surface of this subject, when we were together recently in Davos around the World Economic Forum Annual Meeting to discuss the future of education. How can we begin to develop more agility, more nimbleness, in primary and secondary education, before they come to you and before they go to other post-secondary institutions?

Daphne: In this day and age, people have access to so much knowledge, and we demand that they learn again, and again, and again. What we need to teach, starting at an early age, are those critical skills that are going to enable humans to be agile and responsive to the changing world in which they live.

The arts are a great example of how individuals can be taught to adapt. I am on the national board of 'Young Audiences Arts for Learning,' an affiliate network that serves over five million children across the country. Beyond developing talent, arts learning is an iterative way to teach young people to create and collaborate on developing and executing on a concept —to evolve together. That is a significant departure from how education has been delivered in the past, when individual achievement was the single metric for success. The future is collaborative, and in the case of today's learners, this translates into multi-media productions, and applications in the digital arts aided by building capabilities in computer science and coding.

At WorldQuant University, this has translated into an interdisciplinary MSc degree program in Financial Engineering at the intersection of finance, data science, and computer science. These have been traditionally viewed as very separate fields. But it's at the intersection of these subjects that we nurture critical thinkers who stand a chance of future-proofing their careers.

I believe in five years if you want to do financial management, you'll be better off with a financial engineering degree than with an MBA. As everything is further driven by data and technology, building meaningful knowledge in these areas will be a requirement for successful financial management leadership roles and for making solid business decisions in the workplace.

Authors: What was the catalyst for the creation of WorldQuant University?

Daphne: WorldQuant University was started by Igor Tulchinsky, Founder, Chairman, and CEO of WorldQuant. A financial engineer himself, he has seen that talent is equally distributed globally, but educational opportunity certainly is not. He founded WorldQuant University so that talented individuals across the globe would have access to educational programs that would help them advance their careers without the burden of tuition. WorldQuant does not hire WorldQuant University students upon graduation. We are focused on building the next generation of data-driven decision makers.

We believe that equipping people with these skills has an out-sized impact on their communities. In the short term, we're focused on engaging qualified people worldwide and giving them access to these unique advancement opportunities in order to accelerate their careers. We currently have over 1,800 students, in 90 countries, in the Financial Engineering Master's program, and around two thousand in the Data Science module, all of whom attend tuition-free. Our Data Science module equips students with the data science and analytics skills that are critical for the most in-demand jobs today.

Long-term, we want to create a model for education that sets a new standard for delivery. Brick and mortar institutions worked well in the 20th century. Amid growing student debt, the value of a higher education degree is being reassessed. In today's fast paced dynamic economy, it's harder for these legacy institutions to respond to the changing interests of students and demands of industry. Our goal is to develop highly scalable and agile online programs that are built for what the market is looking for and needs. There are huge opportunities for the future at the intersection of education, government, and business.

Authors: And tell us about the infrastructure that was required to launch this, and how do you see that evolving over the next five years?

Daphne: Our interest lies in delivering high-quality education that is aligned with training. The platform content and skills-building aspect can adapt to market needs, and we will add modules to supplement the core interdisciplinary knowledge base, along with electives based on individual interests and career goals.

Traditionally, finance has been a very strong beneficiary of technology innovation. We have an opportunity here to leverage the skills that have been traditionally applied in the finance industry and bring them to other disciplines that would benefit from innovation. Whether in urban planning, healthcare or agriculture, our hope is to supplement our core curriculum with specific business use cases that people can then translate into important work in their respective communities.

Ultimately, the value we bring is that we're building an ecosystem. We live in a world where education will be increasingly judged by outcomes. And so, we are building another channel, essentially a marketplace, in which companies with job opportunities in geographic locations where we have students can post these positions for our graduates to consider. In fact, we are creating an ecosystem that will beget itself. While I care a lot about increasing access to education, I also care a lot about advancing careers and ensuring people continue to have access to the evolving workforce. By bringing together curricula, educators, learners, and employers with accessibility, values, quality, and the trust of a broader community under one ecosystem, we can deliver the entire package.

Authors: Daphne, we applaud the vision and we can only imagine the process of determining what is possible, which as you have established is infinite. When will the first class graduate, or have they already?

Daphne: Yes, we now have 60 geographically diverse graduates from the program. One of our great sources of pride is that we are creating a global community. We have students in Senegal doing group projects with fellow students in Singapore. They are across the globe – in 90 countries, including in the U.S. and across Asia, and 40% of our students are in sub-Saharan Africa. Through various communication outlets such as WhatsApp, students are expressing gratitude for the friends they have made around the world. Learning how others approach problems with their respective regional nuances can only be expansive to all those who experience it.

This peer-to-peer engagement extends to academics where students are sharing their varied domain expertise. Higher education tends to undervalue the knowledge that students bring to their shared learning environment, and the diversity of expertise that they can share. One might have a background in finance, the other a degree in computer science. Helping one another through the two-year program by leveraging their strengths sets a good precedent for the workplace.

Authors: I'm curious about what you see as the future relationship between corporations and educational institutions like WorldQuant University. How can educational institutions like WorldQuant University and others, play a greater role in re-skilling and also preparing for re-skilling going forward?

Daphne: As the nature of work itself becomes more specialized, corporate survival is going to be predicated on whether they have the right team with the right skills at the right time. We already see more corporations working with educational organizations that can help them up-skill their current employees and vet students from non-traditional educational programs.

One of our hopes is to unearth the talent that in the right environment can develop the skills the marketplace needs. I'm confident we are only going to see this approach become increasingly relevant and more prevalent among educational institutions and companies. These corporate-academic partnerships force both sides of the equation to develop new capacity and evolve their offerings at a pace that keeps up with the changes in the market.

An interesting statistic is that about 40 percent of our students are between 30 and 39 years of age. Everyone has a bachelor's degree, but they are early adopters in recognizing that whatever their career, they need to grow their skillset. Our students have degrees in chemistry, applied physics, financial management, hospitality—nearly every area of study. Presumably, they are assembling a basket of skills that will be highly relevant giving them a competitive advantage moving forward. In the future, educational institution rankings will have to consider career outcomes. What happens five years after someone graduates? How does their career impact the communities in which they live? In our data driven world, we can focus more attention on these measurable, long-term outcomes.

Apart from the typical metrics, we assess these outcomes to predict best fits for individuals, governments, educational institutions, and corporations. Preparing for this agility is no longer optional.

Authors: Let's talk about barriers to entry and growth and what you've witnessed over time. Technology knows certainly fewer boundaries today then when you began and with the introduction of 5G in developed and developing countries, geography will no longer be a barrier. The capital resource pool, both financial and strategic advisors, has never been deeper. Specifically, let's discuss women in technology. You've been a champion for that cause for more than 30 years. We are certainly witnessing an increasing number of women in technology leadership and also in change management roles. What do you attribute this to?

Daphne: Like you, I'm encouraged by the progress we've seen. But, you know, we still have a long, long way to go. The progress there is actually attributable to the hard work of women across the globe. Arguably there is growing recognition that leadership, especially in the tech field, doesn't just require technical knowledge, but takes business, human resources, and most importantly, communication skills, traditionally areas in which women have excelled. It's that kind of dynamic approach that is making a difference in industries where women have been chronically underrepresented.

More and more women are also taking center stage in society, culture, and economics. These disruptions are really just a few years old if you think about it. But it's clear we need to keep focusing on representing diversity of perspectives and backgrounds as we advance technologically and societally. I think the best firms understand that. Perhaps most importantly, women are increasingly recognizing their value.

I've been thinking recently that data scientists are really the 21st century sociologists. Interpreting data is really a social science; it's just that we're working with much bigger datasets now. Previously, we ran surveys with 100 people from which we drew massive societal conclusions. We now have the ability to use huge datasets, but that doesn't make the discriminating qualities we need to bring to the analysis any less relevant than they were one hundred years ago. If the social sciences are where women are well represented, we need to bring the tools of data science to where the women are.

Authors: Daphne, you've been supporting throughout your career, the core premise of the trans human code, which is placing people and positive development at the forefront of that humanity/technology relationship. You've been a business leader, a mentor and now a progressive education enabler. In your closing remarks, how will human centric governance, and the principals of the same, factor into your work going forward?

Daphne: Well, I think it's critical that we all embrace these values. Frankly, for me, technology is a means, not an end. Developing technology for its own sake is not all that interesting. What is awesome is the power to improve or augment the human experience in a meaningful way. That's what drives me.

The tech industry today is certainly grappling with that reality, with inventions that have caused great harm because no one paused to reflect on what they were doing and why they were doing it. We know now what happens if you don't prioritize these principles. So it should be clear to anyone involved at the intersection of technology and education. We have a profound responsibility to innovate with clear intention and purpose; we have the ability to do that in a way that's unprecedented and extraordinarily exciting. For me personally, this work it is not a culmination, but an expression of work that I have been doing for decade and that will move ahead on this continuum.

Authors: We certainly applaud the vision of you and the team and we look forward to virtually attending a graduation ceremony at WorldQuant University soon!

Daphne: Thank you both for championing our work and for giving me an opportunity to share WorldQuant University's extraordinary journey. ▲

Conversation with Federico Gutiérrez Zuluaga

Mayor of Medellín, Colombia, engineer and urban security expert,
and agent of community transformation

Authors: The transHuman code was conceived to create a conversation among the innovators, the implementers, and the users of the technologies that are enabling dynamic change across all the elements of our life ecosystem. Our goal was to provide a forum for the advancement of a prosperous and positive relationship between technology and humanity. We are so pleased to be able to engage you in this important conversation.

Federico, you were born in Medellín in 1974, and you witnessed firsthand the transformation of Medellín into the most violent urban center in the world by 1991. Today, you are charged with leading the second transformation of Medellín, into one of the most technologically advanced and inclusive urban centers in the world. When did you realize that your city could experience its next life, a better life, through the application of technology?

Federico: Many people still have the image of a Medellín from more than 20 or 30 years ago, shown by old news, movies, and television series. But that is not the reality of the city today. Medellín is a transformed city, and that pain and those scars from the past have already healed. We have united as a society to make progress with the city. For that reason, we affirm we are a transformed city, not only physically but also socially. The citizens of Medellín have progressively learned the power of innovation to transform everyone's reality.

I am convinced that innovation transforms the lives of people, and in

Medellín we have used technology to generate profound social changes. We were looking for a solution to reach the ones who need it the most, those citizens who live on the slopes. Of course, this requires new technologies, but its impact is measured by how it transforms the lives of the inhabitants, not by its novelty or technological complexity. The cable cars and the platform "Mi Medellín," are some examples that we have developed in the city that have allowed us to solve major challenges (e.g. mobility, citizen participation) through technology.

Authors: How did your educational and work experience influence your desire to be a change agent for the future of Medellín?

Federico: I am a civil engineer by profession. It is clear to me that the effects of science, technology, and innovation on the generation of better living conditions for citizens is very positive. From a very young age, my vocation has been to help people. My professional background and my experience in public service for almost 20 years have prepared me to serve and lead my community with the conviction that our people deserve and should have a better society, a better city for their lives and their families.

Authors: We met in Davos in January 2016. During that inaugural year, you were credited with advancing many of the initiatives previously established. In the past 24 months, however, under your leadership Medellín has accelerated its pace of technological advancement. At the 2019 World Economic Forum in Davos, it was announced that Medellín will host the first Fourth Industrial Revolution Center in all of Latin America. While Medellín won a global award as the most innovative city of the year in 2012, the goal is to become the nation's tech center. Is this, the Fourth Industrial Revolution Center, a validation of that advancement?

Federico: Receiving this recognition was very important for Medellín. It has been a longtime effort of several governments, and especially all citizens. For us, it is a recognition that goes hand-in-hand with citizen participation, and is a direct recognition of the people of Medellín.

When you gain the confidence of people and institutions such as the WEF, you have gained the satisfaction of knowing that one's city is advancing in the right direction. It also means that important institutions such as the IADB and the national government trust in Medellín.

Authors: What are your expectations for the center?

Federico: Making these advanced technologies contribute to improving the quality of life of people in our countries and, of course, in Medellín. On the other hand, we must avoid the potential trap such technologies pose due to ethical dilemmas and the speed they are reaching all cities. They cannot become a new source of social problems and conflicts, which means to make the speed of the Fourth Industrial Revolution technologies become an instrument to create benefits rather than imply new sacrifices. The WEF's Fourth Industrial Revolution centers are public-private partnerships that use the cities that host them as living laboratories to make prototypes and advance on the practical knowledge of what should and should not be done in our cities.

In this scenario, we hope to receive many companies from all around the world and for them to become partners to our center—companies that not only come to teach us, but also to learn from us on how these revolutionary technologies should be used. Medellín represents a model of emerging cities with unique idiosyncrasies, culture, and DNA. It provides a great opportunity to the world to link with our science, technology, and innovation ecosystem through a skilled entity such as Ruta N.

Authors: How will the citizens of Medellín benefit from its installation?

Federico: Undoubtedly, it will have to contribute to improving the quality of life of our citizens. This is an opportunity to attract foreign investment, to improve our entrepreneurial skills, and to develop new entrepreneurship supported by technologies of the Fourth Industrial Revolution. And the most important, to strengthen the talent of our people for the demands of the current productive sector.

Also, our innovation ecosystem will reach a new level of maturity in a few years, and Medellín will be able to make a great step in its objective of becoming a global hub on innovation issues, attracting advanced companies on these topics, creating world-class talent in these Fourth Industrial Revolution technologies.

Authors: The establishment of Smart Medellín, designed to transform the city from a traditional economy to a knowledge economy between 2011–2021, is a dynamic public/private sector initiative that has made innovation a key feature of the city's agenda. What is the importance of aligning the knowledge, capital resources, and influence of corporate, state, academic, and, most importantly, the citizens?

Federico: For more than 20 years, our city and its governments have made decisions on big issues, working as a team with the academia, private sector, and citizens. This joint work has ensured a clear path for the city, and for the changes to be deep and permanent.

The regions of the world that are being successful in the era of knowledge are not those who have more "smart" people, but those who have many more "average" people connected by innovation initiatives. That is the power of the network, the power of the ecosystem.

Authors: How do you keep the citizens engaged with the Smart Medellín mandate? How do you express the challenges?

Federico: Our city has a long history of integrated work between the public and the private sectors. The citizens who want to put their energy and work at the city's disposal are being taken into account so we can achieve the transformation that society demands. As the leader of the city of Medellín, I am proud to count on that.

Authors: The city has developed many dynamic platforms, such as Ruta N, which is investing in local venture initiatives. Others include the Government Laboratory, Mi Medellín, Cities for Life, Open Data, Medellín Digital, and Comuna Innovates. Tell us about these initiatives, how they are engaging Medellín's stakeholders, and the results that you are witnessing today.

Federico: Two of the initiatives are Medata and Mi Medellín. Medata is an Open Data Strategy that will allow us to be an intelligent city achieving the appropriation, opening, and use of data as a tool of government, citizen action, and decision-making. Currently, we have more than 230 data sets with information from different departments of city government, as well as external sources.

Mi Medellín is a co-creation platform where ideas and inspiration of all citizens come together as part of the transformation of our city involving all the inhabitants of the neighborhoods, representatives of the private sector, and the academy. Since its creation, the platform has received more than 18,000 ideas.

Authors: What is your government's level of commitment to invest in innovation?

Federico: It's very high. Not only in our government, but in our society. Medellín is the city of Colombia that invests the most in science, technology, and innovation activities: 2.14 points of its GDP. Probably one of the highest in the cities of Latin America.

Authors: In its purest form, a community reflects the heart and soul of its citizens. At the center of the transHuman code is our core belief that digital technology can positively transform life for all, as you are clearly demonstrating in Medellín. Do you believe that the success realized to date in Medellín is a result of creating the proper balance between humanity and technology?

Federico: Yes, because we have invested mainly in technology, science, and innovation, but we've been careful to do so only on areas of positive social impact. Our inhabitants are the center of our public policies and our investment. The economy and the developments in science, technology, and innovation are designed for this purpose, to improve the quality of life of all citizens in Medellín.

Authors: From your experience, what guidance and support regarding the importance of putting people first in our future with technological innovation can you offer to other communities around the world?

Federico: Nowadays, a society is not going to move forward producing more of what we have today, unless that city generates not only those elements (which I hope it also does), but also makes sure its central objective is to innovate, to improve the lives of people, giving a priority to find the solution to their most pressing problems as a society.

The great challenges in cities are always based on the needs and knowledge of the territory expressed by its citizens. Therefore, technological transformation is only possible if we ensure it positively solves society's issues and improves the people's quality of life

Authors: Federico, thank you for an insightful and inspiring conversation. We look forward to being together with you again soon as the Center for the Fourth Industrial Revolution in Medellín takes form.

Federico: It will be our pleasure to welcome you here. ▲

Conversation with Marc Deschamps

Executive Chairman of Drake Star Partners, today one of the world's leading technology investment banking firms, founded his first IT company at age 17 before leading major corporate technology initiatives across Europe

Authors: Our goal with the transHuman code was to provide a forum for individuals like you, Marc, to advance a conversation and a dialogue around the relationship between technology and humanity. In the absence of a governor of technology, someone who tells us what we can and can't and should and shouldn't do, our belief is that all of the stakeholders have to be aware of what the impact is.

We are particularly interested in looking at that through your eyes; you've had many lives in technology. So beginning as you did as a student, developing that innovative timing IT protocol for an international racing series when you were just 17 led you to a career in technology management. Your work with market leaders such as Logitech and Philips, you were the COO of British Telecom for a dynamic broadband initiative, developing, leading the first broadband IST with Chello, which was also one of Europe's first unicorns. And now, in your current position, in this third stage of your technology life, you're leading technology M&A transactions, and you've established one of the world's premier technology-focused middle-market investment banks.

You've been witness to such dramatic change in the advancement and adoption of technologies. So we'd like to hear in your own words how you would characterize your experience in this technological evolution and what is it that causes you to jump out of bed every day and keep doing it?

Marc: I would say, first and foremost, it's a deep, strong-rooted belief in innovation itself. Innovation triggers innovation. I have had the chance to experience people and accomplishments that were inspirational, what you could call innovation champions. My dad was a sector champion, and that impacted me. But the core influence for me was that I was put into an environment quickly where I saw a need. Because I was good at math, I was invited into a Formula One car racing event. And at that event, they used about 50 of us students to make the calculations of the racing car data to relay to the teams, the drivers, and the audience. It was the year when the PC came out, so a couple of friends and I decided to put the PC to work on this. That's what triggered my becoming a coder, of becoming a geek, and of believing that technology can transform the world.

Authors: As a student technologist, over 30 years ago, you were certainly an anomaly among your peers. That's changed today. How do you feel our secondary and our postsecondary institutions are faring with their own technology education offerings? Are the institutions themselves providing a solid enough platform—are they encouraging students to the level that they could be, perhaps should be, to become technologically aware if not technologically involved?

Marc: We do need to find ways to bring the kids to technology through experiences that they would like, and that bring the champion influence closer to them. And it's obviously not going to be to make them all coders. That would be a big missed opportunity. It means that we need to bring to them a lot of examples from different people and different vocations. It may be a chef that created a novel innovation or maybe an entrepreneur launching a business service that he created. Give them inspiration through what they already enjoy.

In the formative years of technology development, it was the privileged kids of the universities that were touching some of those innovations. I think we have to engage at a younger age. I think we have to bring those examples of champions to them younger so that they are influenced at an earlier age to determine what they want to do. Doesn't mean that they will know exactly they're going to become the coder, an architect, et cetera at the age of nine. But they can start to express themselves and what they can do, and learn what they can do and know that they can have an influence in the world.

Authors: We, as adults working within corporations, are learning every day the amazing things that can be accomplished with technology, but we can tell you that that's not part of the high school curriculum in most parts of the world. Technology is not working its way into the core curriculum at the primary- and secondary-school level, and it has to.

Marc: I fully agree. I see that my kids are discovering technology at home first. They're discovering the interaction with technology at home on the iPad and with other devices. They're discovering how to interact with a machine, how to interact in communities with their friends within the home environment, which if supported at the home environment can trigger a curiosity that will transcend them and will hopefully travel with them.

So the concept of ideation, the concept of innovation is so important. Our children are already advising their parents and teachers of the importance of clean air. And all for the better because they're going to put pressure on the older generations like me, that we need to move faster in electric vehicles, in consumption management, et cetera.

But at the same time, I want them to understand that they can do more than apply social pressure. Kids can aspire to change the world. That's innovation management, that's coding, that's ideation. The more concepts that the children learn at a young age, the more it will empower them to do those things, I think.

Authors: We were told recently of an entrepreneurship program that is being taught in Singapore beginning at the age of five. This is not within the state education system, but it's being offered as an extracurricular activity, and we understand that there are over 14,000 students registered. So this we think is a great example of introducing creative development skills, perhaps most importantly establishing from the outset that anything is possible.

Marc: I had the great fortune to have an exchange recently with the head of innovations at NASA. They were telling me about their program to move the International Space Station to become a lunar-based station. How to move from bringing people to the moon to now bringing people to Mars. They have a strong belief, of course, that solutions for life on Earth exist beyond our planet. Back to my message of bringing champions forward to our youth, I know that people like this with their ideas, with their technologies, with their innovations, and the financing that I know will come to them can inspire more youth who have access to their messages.

Authors: Marc, we have talked previously about the value and the importance of incubators and accelerators, and we know you've been very supportive of those platforms with your time and capital, across the world. But how do you think that business and finance leaders can make the greatest contribution to these initiatives?

Marc: I think in terms of the innovation financing, let's shift part of the funding to the earliest stage. To the youth. To the primary and secondary education system. For ideation.

I think there's also not enough understanding of what is the importance of the financial markets for innovation. The level of financial literacy, even in developed countries, has to grow exponentially. Money and its management is either understood or it's a frightening subject to a large percentage of the population. As investment bankers, traditionally working for the corporations, we need to adopt the mindset of being enablers for more education that reaches the younger audience.

Authors: The power of the accelerator base unit, the incubator base unit, is reinforcing of the fact that great things come when we collaborate together. Even if we don't know what it is that we're going to collaborate on. Even if we don't know why we're in the same room, the same building, or around the same table as someone, great things can happen. We think the earlier we can introduce that into our educational system, the greater benefit we realize from it.

Marc: Yes, and increasingly, the concept of a team need not be in a physical place anymore. It's the power of the community, community tools, the community links, the messaging, the communications that come in on that, the workplaces, the cloud base, et cetera. So it's how to bring that concept of team to beyond just one class or just a group within that class. How you can link a school that is in the Philippines to a school that is in Montreal, and then another one Paris, that together they can create things?

The team concept, as well, is one where it works best when people have different skill sets. That's a concept that we need to introduce very quickly. It's a concept that we learn in business because I need an accountant on one side, an engineer on the other side, a program manager on the other side, and a team leader to try to put it together, for example. Another benefit from introducing youth to the accelerators and the incubators is that they learn the benefits of collaboration on imagining.

One of the challenges of the corporations, including mine, is we want kids to come in and join us who are ready. It's a fast-paced, unforgiving, competitive market for corporate finance. So we want kids with their fast-paced minds and fast-paced skills to join us. If they are ready, they have an advantage. It is fun and engaging for them, and you can see it in their results. So how can we bring them to be more ready to dive in? I think by creating deeper relations between the corporations and the labs.

Authors: Marc, you really do embody our transHuman code, and you're putting people first in all that you're doing. Not surprisingly, you continue to be successful in your pursuit and application of the technology market knowledge, but also in your financial market knowledge and relationship management expertise.

Marc: I hope I can have more interaction with your transHuman code project and principles supporting it. I think it's a good cause. It's something we have to move on. So thank you very much for the opportunity to contribute.

Authors: Thank you, Marc. This was a very rich conversation. ▲

Conversation with Rodrigo Arboleda

CEO of the Fastrack Institute, cofounder of One Laptop per Child, architect, social innovator, and agent for multilateral exponential transformation of urban centers

Authors: Rodrigo, you've had a most rewarding life with technology, beginning with your study of architecture and the pioneering efforts in the formative stages of the MIT media lab, through today where you've become a leading innovator in transformation through megascale innovation partnerships combining public, private, and even philanthropic collaborations. During this dynamic period of business transformation, I'm curious about what your own personal experience has been and how you've evolved individually through this time.

Rodrigo: Thank you. Yes, I'm an architect by profession, and I went to school at MIT in the 1960s. After MIT, I went back to Colombia, practiced architecture, and became the president of the Colombian Society of Architecture in my fifth year in Medellín. While working there, I was drawn into a project to transition architects into the digital age—with my friend and MIT classmate Nicholas Negroponte—in the early 70s under a sponsorship of IBM, the Ford Foundation, and MIT. AutoCAD originated from this project, providing a graphical user interface for architecture, which was a breakthrough in the use of technology for architecture.

In 1973, the government of the United States decided to map 1,500 key buildings of worldwide importance into digital format, and for that, MIT and Nicholas plotted the city of Las Vegas, creating a 360-degree rudimentary camera on a car that mapped the whole city. That is what is today known as Google Maps.

In the early 1980s, the World Center for Informatics, again with Nicholas, began a new development in Paris that aspired to have children of five years of age and up become involved in digital technologies. That was a far-fetched avant garde, leading-edge type of approach to education. I returned to Colombia, and with the support of President Belisario Betancur, we brought a franchise for that project to Colombia for Latin America.

That was my complete conversion into this new methodology of how to use the digital age to do transformation and education of technology. And that became a passion for the rest of my life. When Nicholas started the Media Lab at MIT in 1985, the first initiative was the continuation of the project that we put together in Colombia. This included 14 schools in the rural areas of the Bogota plateau along with a very large room with 250 Apple computers and IBM PCs, and a big super VAX computer from Digital Equipment Cooperation. And there was born my focus to help societies through education and technology.

And from this project came the idea of "One Laptop per Child." With my private sector experience in business development, the project became such a large opportunity to transform the world. We knew how children could have access to knowledge in quantity and quality of similar caliber as the most privileged child in New York, Tokyo, or even Berlin. But in this particular case, we would bring it to the jungles of the Amazon, the mountains in the Andes, and the plains of Africa.

With Nicholas, I became the CEO of the One Laptop per Child program for almost eight years. And children far removed from the luxuries of cities in developed countries learned how to write and create code and put them into action in their daily life through a digital computer. A machine that consumed a very small amount of energy, that could be read in sunlight, and that could be taken home from school.

We ended up delivering three million laptops worth about $800 million as a not-for-profit organization and managing the entire logistics of manufacturing and distribution in a breakthrough social innovation project that had not been seen before.

Authors: It's very interesting, the trajectory that your life followed and how you availed yourself of technologies as they were evolving for the purpose of applying them for a greater good. You were less engaged in the development and application of technology for personal benefit. You recognized the transformative power of digital technology—not only within the developed countries but the developing countries that you experienced in your youth. And you expanded that exponentially with the institution of the One Laptop per Child program going into markets that no one would have dared or dreamed of going previously.

Rodrigo: Yes, and it is that same vision that brought medical technology entrepreneur Dr. Maurice Ferré; Salim Ismail, Singularity University's founding CEO and author of the well-known book called *Exponential Organizations*; and I to form the Fastrack Institute three years ago in Miami. We concluded that if we don't transform the mindset of the public sector officers in charge of governments and regulatory environment, and if we don't transform the minds of academias, regardless of the motivations of the private sector, we will fail as a nation.

So we formed a not-for-profit entity to help societies, in this particular case cities, as the main element of the future exercise in modern countries, to accelerate rather than reject the adoption of exponential technologies, but only if they can bring together positive social impact.

We chose Medellín, Colombia as a focus because, despite being on the verge of disappearing as a social entity because of the drug trafficking, they have found a combination of private sector, public sector, academia, and NGOs under one common goal, which was how to transform the city through innovation science and technology, especially in the social impacted areas.

That has been my involvement with Medellín. And it has proven to be a successful experience for all parties involved. This also led to a project in Bogota, a project in Miami, and now many other countries and cities are asking us to bring this type of methodology to help them transform into the digital and exponential technologies.

Authors: Rodrigo, let's stop there so we can expand specifically on that. What you've reinforced for us is that the origin of your purpose has evolved organically. Not only have you continually placed yourself at the point of convergence between technology and humanity, but you've always looked forward to what the effect could be for others.

And we think in large part that results in projects finding you, opportunities finding you, and individuals finding you, knowing that you are the ultimate social innovation facilitator. That's a great compliment to you.

Rodrigo: There's a well-known book from the 1950s by a professor from Harvard by the name of Everett Hagen that is called *On the Theory of Social Change*, where he discovered that three cities he had found in the world, one in Japan, the other in Burma, and the third one in Colombia, Medellín, became the agents of development and innovation despite their lack of natural elements—no navigable rivers, no extensive agriculture, no sea ports nearby, completely surrounded by mountains, difficulty of access, and so forth.

So that particular type of culture was instrumental when Medellín was confronted with the horrific drug trade, when they were almost at the verge of destroying society. Public sector, private sector, and academia got together and created a not-for-profit entity called Pro Medellín. And out of that sort of partnership they decided that innovation was the banner.

And inspired by President Kennedy in 1962, the city's leadership declared they wanted to make Medellín the center of innovation science and technology. And for that, the entire forces of society needed to be together.

And that has been the key, how to put the three elements of society—the academia, private sector, and public sector—together in an instrument of development that is unprecedented, and it is the model that Medellín has found. And the one that we propose all cities to take.

Authors: The most recent rebirth of Medellín in Colombia is a tremendous story. How are other cities that have identified specific opportunities addressing key development issues, such as transportation in the city of Miami, advancing?

Rodrigo: Part of the methodology that we follow at Fastrack Institute is to identify critical problems that a society faces. Miami recently engaged us for a mobility project because of the density of the city and the growing attraction of people into the area. Miami is caught with this dilemma today, where there are a few areas with high density like downtown Miami, but the remainder of commuters require long transportation on extremely busy major throughways and highways. So one of the elements for consideration was a new mass transit system.

As you well know, because you are both well versed in Chinese affairs, the Chinese are now very much involved in changing the perception that the world has about them of being copycats of somebody else's inventions. They are now proud and scientific oriented with their own particular inventions. And they have developed a very dynamic low-speed and mid-speed Maglev (magnetic levitation) rail technology, recently implemented in the Hunan province. And we believe that could be a breakthrough in the technology for the future in high-density U.S. cities. It would involve a drastic reduction in capex expenditures and a super drastic reduction in maintenance because there are no moving parts, there's no friction; the train cars are all floating on about 18 mm of air because of the magnetic levitation technology. And we are now engaged in this transformative type of future technology for all public rail services.

Authors: And how will this influence future partnerships for the city of Miami, and how can other cities benefit from the vision and the execution of this initiative?

Rodrigo: Well, we believe that in the future, governments alone cannot solve the problems of society. Private sector alone cannot solve all the problems of society. NGOs and academia can only support and promote ideas about the problems of society. Only the joint venture of these three elements of society—public, private, and academia—under a single objective can do that. And I think that Miami could be the most immediate experiment that this indeed is the way in the future. Because otherwise, it's going to be a complete failure.

Governments alone cannot do it. They would require the creativity, the flexibility, the entrepreneurship of the private sector. And private sector alone cannot win because some of these investments have to be money-losing endeavors. But the idea is to minimize the money-losing part of the equation, but multiply the transformative nature.

Authors: Rodrigo, you have continued to work with countries all over the world, so you're experienced in governments at all level. Is there a distinguishable difference in the embrace of this human-centric approach to ecological transformation in developing countries versus developed? Can you see a difference?

Rodrigo: Yes, I think that the developing nations are more aware nowadays than the developed nations, to be frank with you. And because they are feeling the pinch in more than one way, they are becoming very sensitive. And especially the young generation is becoming so sensitive about the preservation of the environment. And the social impact that technology could bring to the table. That one element that we need to suppress is so much greed into the equation of technology. But if we can implement the aspect of social transformation more, we could start getting a better equilibrium between the forces that affect the developing of nations today, which are technology, money and success—personal success.

I think that people are starting to realize that unless we all act together to save the planet, our children and grandchildren will have a tough world to live in.

Authors: Yes, and do you anticipate that educational institutions are going to play a greater role in the collaboration between private business and public governments?

Rodrigo: Yes, but the educational institutions will have to transform themselves. We are finding that the educational institutions, as we know them today, are going to be a thing of the past in the not too distant future. The educational institutions that are more flexible and more oriented to treat education as a demand problem rather than a supply problem are the ones that are becoming more aware that what they need to produce is what the world is needing. For example, for AI alone, there's about a one million person deficit in the United States right now for people managing algorithms and all the tools that artificial intelligence demands.

Authors: This brings us to the last question that you've answered in part, and that is, as you continue to guide the transformation and collaboration between governments and companies and foundations, how do you think you can most effectively communicate the message to all of these stakeholders that in the absence of the global governor of technology, that they have to be self-governed and self-managed. And always putting people first?

Rodrigo: We have two approaches: One is a brotherhood of cities that is starting to share common science, together with the idea of changing the mindset of the people responsible for all the living aspects of the city. Today with the advent of 3D printers and the makers lab, the new mantra is not anymore learning by doing but learning by building. If you have an idea, come and build it. And showing the practical result is the best way today to change the mindset of people. Because people are tired of being told in a symposium about fantastic ideas and nothing happens after that.

And the technologies are advancing in such a speedy fashion that you don't have time anymore for talking about it; you have to do it. And that goal should be the main approach that we are following right now. Leading by example and learning by building. So perhaps because I'm an architect, and we were trained in design thinking if you don't do the thinking of the design process, you fail as an architect. But they have now applied the same etymology to many other aspects of life, and therefore if you combine all of these elements of transformation and technologies, then you really could have an impact in the world in a positive fashion, and that's our aspiration.

Authors: Well, we think that's a good goal to live by, and you are obviously still living this by the day. And while it has been many years since you began this journey, you still have the same level of enthusiasm when you talk about all these projects.

Rodrigo: Yes, yes indeed. You need to have passion for these things. And thanks to people like the two of you, who are an inspiring element of this whole movement. It is when you find that sort of reciprocal type of thinking, when you discover the echo in the walls that amplifies the message, that you conclude that we are doing the right thing. ▲

Conversation with Julia Christensen Hughes

Dean of the College of Business and Economics at the University of Guelph, advocate and activist for student-centered transformational leadership, and global innovator merging business and corporate social responsibility study

Authors: Julia, throughout your career you have been focused on the quality of teaching and learning in postsecondary institutions. Before becoming dean of the Business School at the University of Guelph, you were the president of the Society for Teaching and Learning in Higher Education and supported the development of a culture of "student-centered learning" as the university's director of teaching support services. Today, the Guelph MBA program is ranked in the top ten globally for its focus on corporate social responsibility.

Your accomplishments tell us that you have been looking forward to the future of students, and the potential of their impact, throughout your career in educational service. To what do you attribute your human-centric approach to educational leadership?

Julia: I've always believed in the power of education to transform lives, and that education at its best helps us sort out who we are and who we aspire to be. I think education, if well led, can expose us to complex problem posing and solving. And increasingly, these questions align with the most challenging problems in the world today.

Students can learn the power of diverse teams to aid in the resolution of problems, how to resolve multiple and seemingly divergent perspectives, and bring them together in powerful ways. And through processes like this, it provides the opportunity to develop essential skills and knowledge bases that help chart one's course for the future or their desired direction in that moment that once the map, the full map or all the dots are connected, becomes that future direction.

Given that the future is often unknown, these skills have to be generalizable to multiple contacts, and as I said, different kinds of problems. Unfortunately, higher education often falls short of this promise due to deeply entrenched systemic barriers. And then I reference my book, taking stock research on teaching and learning in higher education, and I outline what some of these barriers are and what needs to happen to overcome them.

So, I attribute my human-centric approach to educational leadership to believing passionately in the power of education, but also seeing all the barriers that get in the way. And so, I've committed my professional career to try and overcome those barriers by studying them, writing about them, and bringing out my sword and trying to slay them.

Authors: Julia, you returned to the World Economic Forum in Davos this year to continue a very important dialogue that began in 2018, in support of business as a force for good in the world. What do you believe to be the significance of being recognized as a champion business school, and how you are helping to advance the UN sustainable development goals?

Julia: Being recognized as a champion has been hugely helpful to me in the kind of leadership that I'm trying to provide for our business school at the University of Guelph. It is external validation that we're on the right path. When I became dean 10 years ago and we went through a strategic planning exercise, which resulted in our vision to develop leaders for a sustainable world, while the college was committed, we went through a thorough process with focus groups, with all kinds of stakeholders, employers, and students, alumni, faculty, and staff. And when we landed on this vision of leaders for a sustainable world, it really was with almost unanimity. It was very strongly endorsed. We engaged in a thoughtful process, and I was so proud and, in truth, somewhat relieved that we landed on it because the college was comprised of these disparate long-standing units. But we managed to come up with a common vision and, as I like to describe it, the thread that would weave through all of what we do.

But at the time, not many business schools were talking about this. So we really were an outlier, and we did that deliberately. Where we're situated in Guelph, we're surrounded by renowned, well-funded, long-standing business schools, Ivy among others. So, we knew we had to differentiate ourselves, but we also wanted to do that differentiation authentically, and that is where we landed.

But I know we raised some eyebrows. Some people thought maybe this was a passing fad or that we were being too focused on an idiosyncratic issue or something at the margins and not mainstream. So much depends on your personal credibility when you're trying to provide leadership for something, and to have endorsements external to your organization from powerful others are so helpful both in terms of your own morale or spirit core, right? It's important to get that endorsement when you really are working outside the box. And it brings assurance to people who need to be assured, whether senior leadership in your own institution or potential philanthropists, donors, or alumni who are going to step up and support you. So, having the endorsement, first of all, becoming a signatory, and then being recognized as a leader in that group as a prime champion was hugely gratifying and incredibly helpful in telling our story and saying that we're part of a global movement, a critical global movement.

In terms of working with or helping to advance the UN sustainable development goals, we do it in all kinds of ways, and I like to think through the talent that we're graduating, many students who are so passionate about making this kind of contribution. So, we embed the content, we teach a required course, for example, in corporate social responsibility at the undergraduate level. I believe we're one of few business schools that makes the course in this area required, or has until recently. I should say lots of business schools are getting on board with this now that they're seeing this is mainstream for the future and no longer something on the periphery. But in our course, the students learn about the UN sustainable development goals, and their project is to find a company that is making progress toward one or more of these goals and to write their case story. And then we submit these stories to the AIM2Flourish initiative.

Its mandate is to help business schools teach about the sustainable development goals. And so, they've become this repository for all of these case studies the business school students are writing, and every year they choose one remarkable story for each of the 17 SDGs. And I'm proud to say that in the couple of years they've been doing this, every year a team from Guelph has been chosen, which is remarkable; it shows that we're part of a larger ecosystem. That there are businesses in our region that are doing this work, and again, doing it authentically and successfully, that we've been able to expose our students to this kind of thinking so they understand that this really is happening. And again, we've received external recognition in a variety of ways. It's allowed us to send teams of students to the United Nations in New York City, which has been an incredible opportunity for some of our student leaders.

Another example is our hub incubator program. I believe strongly in entrepreneurship and talk about it being a means by which families and communities can become self-determining. I had the opportunity to experience sustainable tourism in Nepal and the Himalayas. When sustainable tourism is done right, when the local people are able to benefit from it financially and culturally, when it's done in a respectful way, that's the power of business to transform lives for the better. All of our students participating in our hub incubator program are coming up with new business ideas. We run them through a mini B Corp certification process so they understand fully what a business looks like that has aligned all of it systems against this direction of business as a force for good.

Authors: Julia, let's talk about the business school ranking initiative that was recently launched to evaluate schools based on the alignment of education with human skills.

Julia: I was invited to facilitate a conversation among deans of champion business schools as well as heads of corporations on the age-old ranking system for business schools. Three questions were posed. One was why do we care about rankings anyway? Who did they serve? So we conducted a stakeholder analysis. The second question was what do we dislike about the rankings as they're currently conceived? And number three, if we could imagine a future desired state, what would it look like? And in what ways would the rankings change to reflect that?

To bring about transformative change, you have to understand what systemic barriers are preventing further progress from being made. And it quickly became apparent to me that the way most rankings work provides incredible incentives, perverse incentives for moving in the opposite direction of business as a force for good. We talk about business school rankings, but if you look carefully, typically what has been ranked is the MBA program. The majority of students studying business are not in MBA programs; they're in undergraduate programs, but with all the focus on the MBA. This can redirect resources away from a high-quality undergraduate business degree experience.

The second thing is that many rankings have put undue emphasis on the salary differential of the students coming into the program and those graduating. This has resulted in all kinds of game playing in terms of what students get accepted and the emphasis of programs. So, many programs have increased their focus on finance to graduate investment bankers who are likely to see the most significant increase in their compensation. And I think in part this action is influenced by a desire to support the current fee structure for MBAs in "premier" schools. But I don't know that this is a metric of quality.

In the current ranking model, the MBA students who join our program to leave the corporate sector and apply their business and leadership skills to not-for-profits, for example, will likely see their salaries go down. Or what about schools like us that put a lot of emphasis in a masters of leadership program that are providing leaders for the uniform services like military, policing, and even hospital administrators? These functions don't show up anywhere on the ranking's performance measurement scale. So that's a big problem.

And third, so much of the weight is based on reputation of schools ranked with outdated criteria. Nowhere in the rankings are undergraduate student voices, nor business minors for non–business school students, and not-for-profit leadership does not carry any weight unless the salaries are comparable to corporate compensation. So, a lot needs to change to put the focus on developing the leaders that the world needs, and I think the student voice has to be key to that.

Fortunately, the *Financial Times* has announced their willingness to do a fundamental rethink. What they don't support is interdisciplinarity, and a lot of the work that needs to be done to support the UN sustainable development goals, for example, happens at the margins. It's the magic when different disciplines come together and creatively solve problems, drawing on multiple perspectives. Journals that do that or books that do that don't rank; they don't even count.

Authors: The University of Guelph was an early adoptor of the practice of cooperative education. What do you think is the greatest challenge for corporations to prepare for changing work roles? And would you expect that educational organizations and companies are going to collaborate more closely going forward?

Julia: Our pillars are active learning, research with impact, and community engagement. And I often say only by getting the last one right will everything else happen. I think this notion of sending a child away to school for four or five years, I think that's an incredible luxury. I think it will continue to happen for families of wealth, but I would hope that even in those situations, that students have the opportunity to engage in co-op placement throughout. We know that when students get that real-world experience to develop skills, they bring their learning back into the classroom for their subsequent courses and they get so much more out of it. I do think that's the future.

I also think there will be much more novel arrangements. I think we're going to see increasing alignment between the business world and secondary/postsecondary institutions that's going to provide much more accessible, affordable education with that applied focus.

Authors: And how will incubators and accelerators factor in that development process, do you believe? Are we just in the early stages, do you think, of that model?

Julia: Yes. This is going to continue to grow. I cannot believe what has happened in such a short few years since we've launched our incubator at Guelph. But what we're finding is that we want to provide multiple ways for students to learn entrepreneurship and innovation skills. And it's everything from noncredit start-up weekends—if you've got an idea, come and brainstorm, come and learn about the ideation process—to offering a minor in entrepreneurship that would be available to any student at the university. We find so many creative ideas coming out of the biological sciences, computer science, engineering. How do we help those students take courses in business and entrepreneurship is a key question that we are asking.

And then we have the incubators for the students whose ideas are further along. To them, we can support with financial resources, coaching, taking them through to launch and then helping them accelerate. And we are now witness to student-run businesses already exporting product around the world in just a couple of years. It's amazing.

And for students who don't go on to launch their own businesses, they'll bring an innovative or an entrepreneurial mindset to corporations that they join, always asking themselves, *How can we do this better?* And I think that's to the benefit of everybody.

Authors: Julia, you truly embody the core principles and the transHuman code as we've envisioned it, and we are appreciative of the opportunity to be on this journey with you.

Julia: Thank you for this important initiative and allowing me to contribute to it. ▲

Conversation with David Shrier

CEO of Distilled Identity, globally recognized authority on financial innovation, government advisor on economic expansion, author of multiple books on fintech, blockchain, and cybersecurity, with dual appointments at MIT and Oxford

Authors: David, you're unquestionably a pioneer. You've become a globally recognized authority on financial innovation. And for decades, you've been working with corporations and governments to generate economic expansion. You hold academic appointments with two of the world's leading educational institutions, MIT and Oxford. We are interested to begin this conversation by talking about what you've been witness to through this transformative period of digital innovation.

David: I've been fascinated throughout my career with this question of how technology can solve big problems. And it's been interesting to play that out in both academia and the corporate life. On one hand, I'm building businesses and working with large companies, as well as running startups and seeing how you create innovation from the ground up. As well, I've been privileged to be able, for the last 19 years, to sit within the halls of academia and work with the next generation of innovators. I get to learn a lot from my students.

I knew nothing about blockchain until some students came to me and said teach us about blockchain. And so I thought this was a good opportunity to learn. But I had been a database programmer forever and worked in data companies. And blockchain is, basically, a database. What I find is that there are a lot of different stakeholders in an ecosystem who are trying to move things in a new direction. You get an interplay between government and corporates and academia. I think that intersection is not well understood. Just this year, I have initiated a multiyear project to learn more about how we engineer successful innovation ecosystems built around this theme of technology transformation.

Authors: It's the pace of change that we find most daunting. The financial services sector, in particular, has always been one of the core pillars of market economies. So you're going to become a student, and hopefully a teacher, of how government, companies, and educational institutions can converge and be most effective. Is the regulatory infrastructure in developed markets like the UK, EU, and US capable of withstanding the pace of change as we see it now?

David: That's a really interesting question because, in fact, particularly over the last six years I've been spending a lot of time with regulators and policymakers who are trying to figure out how to grapple with disruptive change. To your point, the pace of change in financial innovation is accelerating. Historically, regulation tends to lag innovation, and that is certainly the attitude I've heard expressed by a number of the regulators and the policymakers in developed markets like the US and the UK.

Increasingly, we're starting to see, in pockets, regulatory activism. It's been more common in the emerging markets, so in places like the Emirates, or Bermuda, or Mauritius you see the governments stepping forward and trying to stimulate innovation through direct government policy in the financial sector. Whereas in the US or UK, there's been a cautious assessment and then a reaction to whatever the technology innovation is, and that's deliberate.

We spend quite a bit of time putting together some analysis of this, and there's almost a Hippocratic oath of policy—you know, first do no harm. When you misstep, you get something like the BitLicense in New

York state, which had the effect of, basically, benching New York City as an innovation center for a couple of years until people figured out a different approach.

The short answer is you usually don't want the regulators leading because that can lead to stillborn innovation. You'll end up with a government infrastructure that supports going one direction and the market wants to go in a different direction. That said, the regulators and policymakers are seeking to get a bit more responsive as they recognize this rapid change in what's going on. For example, entities like FINRA, the self-regulatory body within the US under the SEC, has formed a FinTech advisory committee. I'm one of the members. And they're doing that so they can get smarter faster about evolving changes in the securities landscape.

The Department for International Trade in the UK has formed a FinTech committee with the city of London, again, to become more responsive to industry needs. This just follows on the UK government creating Innovate Finance UK, which itself was an effort to bring government and industry closer together. You'll notice in this analysis that I have not been talking much about the EU, and that's for a reason—the EU is in the middle of a really interesting and ambitious experiment.

I've been working with the EU in one way or another, and the European Commission, for about a decade. The older model, or the prior model, had been similar to other governments in the G7 or the developed world, where they'd be watching new things emerge and looking to generate a regulatory response to it. And their views around AI, for example, are following that path. But they're trying a new experiment with blockchain, which is kind of interesting, because they're going a step further than the typical government R&D support in setting up their

300 million euro blockchain fund and really trying to use blockchain as a way to address structural deficits in how new businesses get formed in the EU and how they get funded, their capital formation and growth. It'll be interesting to see how that experiment plays out.

Finally, the interesting thing that we're starting to see is the OECD, which you don't historically think of as a bleeding-edge innovative body, has started to bring in new leadership and take efforts to drive the conversation rather than respond to the conversation. So across 35 member nations plus their affiliates, particularly the financial directorate under Greg Medcraft, they are seeking to promote new standards and harmonize action around engaging with cutting-edge technologies like blockchain and artificial intelligence.

Authors: David, what is the composition of the OECD? Is it principally or exclusively developed countries?

David: Yeah, so the way to think about the OECD, and I'm not going to speak in an official capacity on behalf of the OECD, but just to provide some perspective, is comprised of 36 member countries that are principally the most developed countries in the world. So it's North America, Western Europe, and then Turkey, and Greece, as well as Southern Europe and Turkey are members. Japan is another member state, so it's not officially a founding member state, but it's another member state along with Australia and Mexico.

But you can think of it as the 36 or so most developed economies, and emerging economies are to participate in the conversations and be a part of the discussion.

Authors: Who is going to bring the greatest value to the developing countries, and what will that look like? How can educational institutions and how can corporations bring value without irresponsibly capitalizing on those markets that are looking for the most support technologically, financially, socially, philanthropically?

David: That's an interesting question. The World Bank, the IFC, and other organizations like DFID, which is the UK development fund, have been working in the emerging economies for years. But we're also seeing the emergence of private-sector actors who are looking to promote technology-led development in emerging economies separate from direct government action or the action of nongovernmental, or philanthropic, or multigovernmental bodies.

I ran into a group that is talking about economic developmentals versus sustainable developmentals and saying, essentially, we're going to teach countries how to develop their own economies on a for-profit basis and to make the World Bank irrelevant instead of having them depend on handouts. I think what's going to be most effective is too early to tell. I'm guessing we're going to see a variegated landscape play out where these various bodies will be trying out different interventions to see what works. And it's interesting that for all that they're being vilified in the Western markets, in the US, Canada, et cetera, Huawei is doing a lot of positive good for development in Africa.

And while they may be financially motivated, it's also helping our least developed continent on the planet with critical digital inclusion.

Authors: Investments by China and by India into the infrastructure across sub-Saharan Africa, in large part, is proving to be an engine for growth. And whilst growth has typically been slower as a result of legacy institutions in developed markets, several of those legacy institutions are actually leading the revolution in some of the developing countries.

David: Yeah, that's a really interesting point. I run an AI-driven start-up company, and, among other things, one of our core areas of focus is promoting identity inclusion and digital inclusion. And what we've found in seeking partners in that effort that is some of the so-called digital innovators, I'm not going to name names, but companies you would think of in FinTech as being progressive or forward-leaning (*think of* being the operative words), have proven to be resistant to engaging. They're sitting back and saying, well, show me ten other people who are doing this and then we'll work with you. Whereas some of the names you would think of as being very traditional and slow-moving incumbents are proving to be the most engaged and leaning forward the most.

You get this when you have innovation inflection points with markets in transition. We're going through another transition of things like payments, banking, and current accounts. A lot of the core of retail and small business infrastructure is getting reshaped right now. And the responses show that we've got a market in transition. It's going to be very interesting to see how it plays out over the next three to five years.

Authors: Yes, well, to say it's dynamic is an understatement. That's for sure. Right here at home, and certainly all across Europe, we're on the verge of what the statistics would tell us is an HR dilemma. By 2022, the World Economic Forum's most recent jobs report tells us that 75 million jobs are going to be replaced by AI and machine learning. Yet 135 million individuals are going to be required to fill work roles that don't exist today. It's a staggering number.

What does the declaration of that chasm between 75 million and 135 million and the purpose for it say to you? What thoughts does it evoke?

David: First of all, I believe the 75 million figure. I'm less confident in the magnitude of the jobs creation figure. But let's pretend we have a growing population, so as a percentage of population, maybe it's believable. The kind of disruption we are seeing in the labor force is comparable to what we saw with the industrial revolution. If you think about that, you had factories that had a lot of manual labor, and then the steam engine came along and suddenly you were able to completely reshape how things were produced. And entire categories of jobs became obsolete. And so it's that level of disruption.

That's just happening. We're getting this massive productivity improvement through the introduction of smarter thinking machines and integration of them into various industry sectors. But innovation doesn't just have to happen to us; we can do something about our response to it. That's where we get into this question of rescaling the labor force and being smart about it. We were not smart about how we handled automation-related labor displacement in the 80s and 90s with the steelworkers in the British Midlands or the auto workers in the US heartland. And as a result, the companies, basically, optimized profit. They did what they were incentivized to do, and they fired people. And it was left to "society" to deal with the consequences.

Versus, say, how the US military handles demobilizing a soldier and bringing them back into society where they're given training, they're given educational credit, they're given support for transition. And there's a much more structured intelligent approach of what we do with someone when we say okay, your job is no longer X, you have to figure out what your new job is.

And so I'm hopeful, more than just hopeful, I'm advocating for an active effort to re-skill these workers, and to create that new skilled worker of the future that this report talks about in terms of the need for people with new capabilities around these new jobs. And part of that is going to require greater cognitive flexibility. One of the companies that is leading this charge with AI is called RIF learning, and it's part of the broader theme of computational social science. But the idea is that we can work with AI to teach the brain to become better at acquiring new knowledge faster.

And that's going to become important because, with the pace of technology change accelerating, we're going to have to get a new certification or learn a new skill every one to two years. And so that's going to require a different kind of education than the conventional four-year degree. Instead maybe you get a master's degree, and then you work for the rest of your life. And maybe you get a corporate training every three, four, or five years for some short course, but that's about it. So now, we're going to need to actually teach people to be more nimble and help reform how they acquire knowledge. And AI can help do that.

Authors: David, would you envision the opportunity for individuals, or the encouragement of individuals, to remain attached to their institutions, to subscribe, for lack of a better word?

David: Do you mean their academic institutions?

Authors: Yes, their academic institutions.

David: Potentially. I have to be careful about talking my own book because, you know, I do have a couple of institutions that I'm affiliated with. We do offer digital classes where we try and re-skill people. So with self-interest declared up front, let me argue that it's going to be helpful because you know the brand and you know the brand of thought, the model of learning. So the model of learning at MIT is different than the model of learning at, say, UPenn or even Harvard. And so if you're used to the MIT brand of education, then you're going to have an affinity for that on an ongoing basis. Even down to the level of when we were playing around with launching these online classes, we were struggling with the grading heuristic. When we tried an experiment and made one of the modules a little easier, we got complaints because students said, I took this because I expected it to be MIT hard.

Authors: Very interesting!

David: We actually had to go back and make it more difficult. And then when we set up the classes at Oxford, we consciously applied the European grading rubric, which is a lot tougher than the US grading rubric. US education tends to have a consumer-oriented, almost entertainment-style brand where everyone expects to get an A for effort. And in Europe, I find the grading standards tougher. We had to make that adjustment because the brand for Oxford was different than any of the US brands that I'm familiar with.

I do think people are going to have some affinity, but the other opportunity that digital technology provides us is that you can window shop and comparison shop. And you can mix and match, so you could get your . . . they're going to kill me for this, but you get your strategy class from Harvard Business School, and you can get your analytics class from MIT Sloan.

You'll pick the best of each institution in crafting your portfolio of knowledge.

Authors: Very interesting. So it's a personally curated education plan because it doesn't just end with three, four, or the first seven years of education.

David: That's really an astute turn of phrase, a personally curated education plan, because you're not going to be able to just have someone hand you the curriculum. Yes, you're going to have to design your career and design your learning to match your skills and your opportunities. It's going to be more complicated to deal with in the future, but this is where AI also can help. You can almost imagine you'd have a robo advisor, but for your career. That's something that could emerge over time.

Authors: That sounds fascinating. We are going to move a few steps back from there. Let's discuss primary and secondary education. What can happen and what needs to happen there? What do you believe, as an academic working in premier educational institutions, should happen?

David: Wow, yeah. Now we're getting into an area where I will speak from a reservoir of vast ignorance because I am not a primary educator. I am a father of two children, now ages 10 and 12. I watch them go through the educational process, but I want to emphasize that although I've worked around the field, I'm not someone who does primary education. I've been teaching only at the university level for the last 19 years.

So with that caveat, I think it's all broken. The way we teach young people is ridiculously bad. We've gotten into . . . certainly in the US, and I'm going to speak now from the US perspective because it's done very differently in other places around the world. In the US model, there's too much emphasis on test taking. There's too much emphasis on memorization and regurgitation. There's a loss of critical thinking skills. The academic system encourages conformity because it is, basically, easier to scale. It's more cost effective and easier for the teachers. The concept of tenure for a primary school teacher leads to a tremendous amount of mediocrity because it's not held to the same standards of tenure that, let's say, MIT or Harvard is. And it's underfunded, and we have antiscience forces injecting opinion and non-fact-based dogma into academic textbooks.

We've got a lot of problems in the educational system. And I'm not quite sure how we're going to fix it. You get people who come through this very regimented memorization, regurgitation, standardized test mode, and then they graduate from high school. And you throw some of them into excellent universities, which are much more unstructured in terms of how you're supposed to develop critical thinking skills and form your own opinions and generate knowledge. And some of these students have floundered because they haven't been prepared for it. And worse, only those privileged few at the very top of the pyramid are able to avail themselves of these more dynamic learning environments. And some of the big state schools, and community colleges, and online universities tend more toward that memorization and regurgitation model. So it's starting to bleed upward, too, into university education.

We're setting them up for failure in the workforce because the employers of today and tomorrow want creativity, flexibility, critical thinking, cognitive dynamism. And nothing that we do up until university, if we're lucky, is preparing those students for that kind of environment.

I know when we started putting classes online, we said we're going to do it different because the online model is starting to repeat the worst sins of the conventional mode. So we're going to inject principles of cognitive science and what we've learned about the brain over the last 50 years into the design of the course. And we're going to have it be project-based and team-based and teach critical thinking. Even though it's harder to grade, we're going to set up a teaching team to be able to handle grading freeform or user-generated content as opposed to relying exclusively on standardized tests. It's been transformational for the now 10,000 students that we've put through this model. They're in 120 countries, and they're changing the world.

Authors: We're curious about your outlook on how educational organizations, institutions, and companies could, or may have to, collaborate more closely going forward. How are you envisioning institutions like MIT and Oxford playing a role in developing re-skilling programs, implementing re-skilling programs, perhaps even grading re-skilling programs? We are imagining that there's going to be a more enhanced role for educational institutions to play in helping corporations reshape their workforce.

David: I think that's right. I've been trying a few different experiments over the last few years on how to do that, both at MIT and at Oxford. We had this FinTech entrepreneurship class we taught on campus called

Future Commerce. It was the first graduate FinTech class in North America. Students wanted to go a step further, so we created this on-campus accelerator program to help bring the ideas forward to the point where they were ready to turn into businesses. And a number of those ideas came off of the back of academic research. So that's one model.

In another model that we are developing, we're working with a large financial services company in Europe with the University of Oxford at the Saïd Business School around a program specifically targeted around AI for corporate innovations. We would be collaborating with them on developing educational content, on the one hand, that would help not only their own employees but others get smarter about AI and translate their skills from other domains into AI. But also, it could be used to tune up specific problems within this corporation that they're trying to solve and get different teams of students working on those problems.

The benefits to the corporate partner are that they're getting ideas, they're getting people solving critical problems for the business, they're getting a pipeline of future employees, and they're getting to expand their brand footprint as an innovative organization from a student standpoint and the academic institution standpoint. So from a student standpoint, the students are getting hands-on experience with real-world problems married to sound tested academic frameworks or strategy frameworks that draw on the knowledge from hundreds of companies in order to formulate best practices of innovation. In this case, innovation around AI.

And then, from the academic institution standpoint, this is an opportunity not only to improve the quality of learning, the quality of pedagogy that Oxford is able to deliver by tying it even more closely to hands-on or application of knowledge to practice. But also, it creates a dynamic test bed for faculty research. Academic researchers are constantly looking for more and better data. All too often, academic studies are conducted on groups of first-year university students because that's the available pool. The more that we can get data drawn from real-world environments, the better the quality of thought that's going to emerge from academic institutions.

And this is something I've been working on, pieces of this model for two decades. This is going to be my first chance to bring all of them together on how to closely link the most abstruse areas of academic thought with the most applied areas of corporate reality and have a seamless migration from one to the other.

Authors: Well, we applaud your vision and your persistence for sticking with it for 20 years. Its going to be very inspiring to other academic institutions, and we can envision the current business model of the university or college today becoming transformed over the next couple of decades.

David: I couldn't agree more. And we're actually going to stand up another company this year focused on that end of the spectrum of innovating the educational experience in a more corporate-friendly, or real-world friendly, fashion, which will help all the stakeholders.

Authors: In your role as an innovation steward, an academic leader, a guide to corporations and governments, we are interested to know, how does human-centric technology development factor into your thought process, and how is it guiding your day-to-day activities?

David: I spend a good deal of time on this question of how we shape AI rather than having it shape us. And one key example is around the ethics of AI. When we build a new AI algorithm, what are its consequences and unintended consequences? Automated lending seems like a good idea. Now we're going to be able to make more decisions and make more loans to more people. But if you let it go unregulated, the AI algorithms will teach themselves to be discriminatory. We begin to automate inequality, which happens to be the name of the book that talks about this algorithmic discrimination.

Another example is around, basically, what's happening with news and media where the Facebook newsfeed, and Facebook and Twitter, and the social media platforms have been actively used by state actors to reshape thought of the electorate and change the course of Western democracy. I say this without exaggeration. I was just reading a quote this morning where one of the senior intelligence officials said that, and you can probably figure out which country, was bragging about how they managed to change the way people think through this application of AI. And it was an AI human hybrid system.

As a business innovation example, it was very cutting edge with this mix of people and algorithms that were generating fake news and then strategically deploying it to particular influencers within the social network and then reinforcing it and creating bubbles of fake news. But the effects are horrifying. I've been working to inject more ethics into the AI discussion. We, human society, are building it, and so we, human society, can decide what we apply this new technology to.

Authors: David, we really appreciate your sharing of experience and insight. And thank you for allowing us to look into your future of transformational innovation. We look forward to continuing this dialogue.

David: It was my pleasure to contribute to the transHuman code conversation. These are important subjects to discuss for our collective future. ▲

Conversation with Risto Siilasmaa

Chairman of the Board of Directors of Nokia, cybersecurity
pioneer, and author of *Transforming Nokia*

Authors: Risto, you're a pioneer in cybersecurity, credited with one of
the world's foremost technology company turnarounds, and today you
are a driving force behind innovation during this digital revolution. And
most importantly and most interestingly for us, you're leading this charge
with early-stage, mid-stage, and, in the case of Nokia with 150 years of
history, mature-stage companies.

The transHuman code was envisioned to create a dialogue, a conversation among the innovators, the implementers, and the users of technology that are enabling such dynamic change across all the elements of our
life ecosystem. Our goal was to provide a forum for the advancement of a
prosperous and positive relationship between technology and humanity,
and we're pleased to be able to engage you in what we believe to be an
extremely important conversation.

You're widely credited with Nokia's advancement from a struggling
company that had previously been at the forefront of the digital era before
succumbing to market pressures, into what the media heralds as one of the
most successful transformations ever. Now with Nokia as a firmly established global technology industry leader, what are some of the core values
that you attribute this turnaround to?

Risto: Well, as always there's never a single thing or value or process that
enables successful transformation, but one aspect that we put to practice
that didn't exist in the company before was scenario-based thinking. Let
me put that into a larger context. Now with data science taking long steps
forward with machine learning and all sorts of new technologies that

enable us to make sense or at least create statistical theories out of huge amounts of data, we face a situation where a leader may be confronted by an unwelcome interpretation of the company's future pretty often because there's so much data and there's so much machinery that can analyze that data and form conclusions.

And if the leader's gut reaction is, "You can't come here and predict that our business has no basis going forward. You can't do that. Go away. Come back when you have real answers," this is unfortunately still a reaction that many leaders have when confronted with things they don't like. So instead, in this new world, where we'll have these situations much more often, the reaction should be, "Okay. That's one scenario. I'm not saying I believe in it. I'm not saying that I don't believe in it. I'm not saying that I like it or I dislike it. It's just one scenario that we need to put some effort into to understand better."

And if we can create a leadership culture where bad news is welcomed, or let's say at least it is not unwelcome because we don't need to buy into the bad news, we'll just say, "Okay. This creates a new scenario for us." As no company has resources to track an unlimited number of scenarios, we have to pick a limited number of plausible ones that we start working on.

For any of the selected scenarios we want to understand how we would know at the earliest possible moment that the scenario is turning into reality? What kind of metrics, what KPIs might add validity to this scenario? What kind of data would disprove it?

And, there are always actions we can take right away. For a negative scenario, is there something we can do that will decrease the likelihood of that scenario happening, something we can do right now or over the next days, weeks, months? Let's change the probability curve for that scenario so it becomes less likely.

If it's a positive one, the opposite: What can we do to increase the probability? And then we have a number of scenarios at different levels of the company, and that enables us to deal with uncertainty and to create a culture where data is just data. It doesn't have a color. It's easy to see how we will have to deal with more data going forward. And this is what we did at Nokia.

We had a huge number of scenarios, some of which led to the disappearance of the company, and some of which led to the transformation of the company. And we dealt with all those scenarios in a systematic, data-driven, analytical way, both in the board and in the management team. And then we rinsed and repeated that process, gathering new data, changing the scenarios, killing some, creating new ones, and we navigated our way through a pretty wild ride.

Authors: That's truly fascinating. What was the catalyst for this event? Was there one particular moment, or did this evolve over time from what you were witnessing in your early-stage role?

Risto: Well, this is sort of based on my way of thinking about strategy. An entrepreneurial way of thinking about the world, at least an engineering-oriented, entrepreneurial way. But what really started it at Nokia was Microsoft's announcement of the Surface tablet. I don't know how well you know Microsoft, but everybody has always thought about Microsoft as a software company, and their partnerships have been key to Microsoft's success.

They do the operating system, and they sell it to their OEMs, the Dells and the Compaqs, HPs of the world, and those companies have paid license fees to Microsoft. And then Microsoft's business model evolved, and they started competing directly with those OEMs by bringing their own laptop, the Surface tablet, to the market. It was such an earthshaking moment that it would have been silly for us to ignore it, and therefore we started thinking, "Okay. If Microsoft starts competing with those companies, why wouldn't they start competing with us and bring their own smartphones into the market?"And we were joined at the hip with Microsoft in an exclusive relationship, where we were tied to their platform: the Windows phone operating system.

And they could become a competitor, so naturally we said, "Okay. What if that happens? What would we do then? How do we prevent it? What else could happen?" And then we started down this scenario-planning path. And we had a huge number of scenarios for each one of our businesses. And it was an extremely educational process. When we decided to sell our mobile handset business to Microsoft, despite that being a heart-wrenching moment—we were selling the crown jewels of Finland to a company that was thought by some to be an adversary, but we knew that it was the right thing to do—there was no hesitation from our side because of all of this exhaustive, scenario-based analysis. We had considered every other alternative, and we knew that this was significantly better than the second best, so it was an easy decision for us to do. What remained was of course convincing the rest of the world that this was the best way forward.

Authors: We are firm believers in the theory called SynchroDestiny. At the core of Deepak Chopra's study and teachings of the science of coincidence are those incidents that are intended to align with one another. We often hear people use the term, "Everything happens for a reason," but obviously in this situation you had preconceived of this approach to decision making and transformation.

And you were given an opportunity to be able to apply that at the absolute grandest scale for a company that was the pride of Finland. And I'm sure that the collaboration with other companies at varying stages through the process were carefully evaluated, considered, on some occasions challenged, so we can only imagine the volume and the complexity of discussion that existed internally within your company before you made the decision to bring it forward.

Risto: I like to think that we can influence those seemingly random occurrences. Like Jack Nicklaus, when being complimented for his 18th hole in one by someone saying, "Congratulations, what a lucky shot!" And he answered, "Thanks, I have noticed that the more I practice the luckier I get." This is what scenario-based thinking can bring to a company. We end up influencing our own future, not just the future described in the official business plan but also the futures we do not want and the futures we do not believe in. We shift the possibility distribution to our advantage and create the perception of being really lucky.

Authors: Risto, We are also particularly intrigued with the Future X Network initiative. At the core of that is innovation through collaboration, and as that is one of the key pillars of our transHuman code, we'd like to learn more about the vision for the Future X Network, some of the activities to date, as well as show you're measuring success through that program.

Risto: One of the things that we obviously are doing is pushing the capabilities of networks and especially wireless networks to a completely new level. But why are we doing that? Why is it meaningful? Perhaps the most meaningful reason there is is to solve the Solow Paradox.

So why aren't we seeing the kind of productivity growth that has been there for the previous industrial revolutions? The fact apparently is that we have digitized roughly 30 percent of value creation in the world, and for that 30 percent the productivity growth has been roughly 2.8 percent annually, which is exactly where it should be historically speaking, but for the 70 percent of value creation that has not been digitized, productivity growth has been a little bit less than a percentage point.

But now, with the new capabilities of networking, which are

necessary for real industrial automation, real remotely, wirelessly managed robots, autonomous driving, and automated logistic centers, for example, we can actually start digitizing the part of value creation that has not been digitized before. And that obviously promises huge amounts of higher value creation.

Like in the United States, Bell Labs has calculated that we may see a sustainable productivity growth of close to 3 percent annually from 2028 onwards. This would result in trillions of dollars on an annual basis in just a few short years. So very meaningful from the global perspective and hopefully also driving us to solve some of the existential issues humankind faces, global warming, of course, being the primary among them.

That's a very high-level reason why we are happy, in the morning, when we wake up, all of us Nokians, to answer the question of why. Why am I getting up? Why am I going to the office? Why is this meaningful? I could talk to you about the technologies, the things that are changing, but I think most people are not that keen on the technicalities of things. This high-level, purpose-driven view is perhaps more meaningful.

Authors: Yes. We appreciate that very much. The transformation of the organization required that everyone within the company adopt the vision that you had curated, for lack of a better word. You're not dictating that vision to the employees, but you've presented it to them to embrace. Through that transformation, did you find that you were enabling the members of the Nokia team to be more considerate of new approaches to business operation?

Risto: Well, as you well know, nothing is ever perfect and everything is a journey, so it would be foolish to claim that every single Nokia employee everywhere in the world sees our vision the same way. We have several cultures in the company. We are built based on a number of acquisitions, so different parts of the company have very different heritages, but we do have a joint mission, and we do have a vision of the world that binds us together.

We call that vision the programmable world, and the Future X Network is the technology that creates that world or that is our contribution for that world. And the programmable world obviously means that we will be gathering an unheard of amount of information about what is happening in the world. And we are building the capabilities. As humankind, we are building the capabilities to make sense out of that volume of data.

And then we can influence the real world back, based on the understanding we have from that data, and that obviously makes the whole world programmable. We can decide in advance what should happen in the real world if A and B happens. And of course with machine learning, it can be A and B and then we run out of characters because so many things can combine, causing a reaction. That's an exciting vision, but it's also a little bit of a scary vision, certainly with cybersecurity issues. If the world is programmable, who will be doing the programming?

Authors: Oh, of course. One of the issues that prompted us from the outset to conceive this platform—that we are without a global governor of technology. The question of "What technologies can be adopted, what should be applied, how should they be managed, how should they be overseen?" is one of the great challenges we face. One of the takeaways for us in this conversation so far is your desire not only to realize the greatest potential for Nokia and all of its stakeholders but also to have a significant impact on people all over the world. Of course, that requires thoughtful development and in many cases the adoption of innovative ways of communicating.

We know that you and the organization are leading contributors to the key policy debates that are happening around the world with regards to connected society and the adoption of new technologies. What are your thoughts around how public and private partnerships can contribute toward the advancement of equality among all people? How are you envisioning the relationship between civic administration and Nokia?

Risto: We have offices in more than 100 countries, but we work in many more because networks are being built everywhere. Our company has a long history of working with governments because we build the most complex networks in the world for local teleoperators to run, and obviously those are part of the critical infrastructure, which brings the government into play. And the government or the nation owns the frequencies that are being used, so there has always been a public/private partnership between the operator and the government and us.

Now, of course, we are seeing more and more reasons for this public/private partnership. And again I'm referring to the existential challenges that humankind faces with pollution, with the availability of fresh water, food for everybody. Poverty is still an issue, although we have made giant strides over the last 20 years. And global warming contributes to all these other challenges and adds new ones, such as refugee issues, large-scale migration.

And that brings us to political instability that is augmented by all these. There are always politicians who want to take advantage of riding on a wave of an external enemy, be that refugees or whatever. So this is really a moment in time, in world history, when we need to be more united than ever to tackle issues that no nation alone can tackle.

And we have a sort of reverse-networking effect in play, where if everybody is not on board, then it's really difficult for a subportion of humanity to make a sacrifice because they feel that it doesn't matter:

"Even if we sacrifice everything, it doesn't move the needle." We all need to move forward in a united fashion, and therefore public/private partnerships, in my opinion, are needed on a different level.

Many large companies are heavily involved in these topics through industry associations, through World Economic Forum, through United Nations, and so forth. And Nokia is very much included on that list, so we feel that we have an obligation to all of humanity.

What's really bothering me is how people think about sustainability and how companies think about sustainability, how investors think about sustainability. They think that a company's or any business's role in sustainability is to show how they use less electricity, how they shut down the lights in the offices for the night, how we reduce our negative impact on the world, and that's, in my opinion, looking at it the wrong way. We should be talking about the net impact a company has overall. If what we do as a business is helpful, surely that outweighs the negative things that happen while we do what we do for the greater good.

And whether we can reduce the negative effects by 5 percent or 3 percent is almost meaningless in the face of our whole operation helping the world survive and tackle these challenges. And then we have certain companies whose whole operation is negative. Their business, their products are adverse to health, for example, and therefore if they are respected because they spent a bit less electricity last year in generating this huge negative net impact, they shouldn't be respected for that.

We should really be calculating the net impact of businesses. And if investors want to be ESG investors, they should not be thinking about the electricity bill or "They spent 3 percent less on rare metals," or whatever. They should be looking at the net impact.

Authors: Yes. We think we are overdue for the development of some new measurement metrics for corporate contribution. And perhaps that can evolve out of more of this type of dialogue. Is that a conversation that you're having with other CEOs today?

Risto: There is an organization that we support called www.uprightproject.com. There is an Upright Project, which is seeking to use machine learning and science to calculate the net impact of large companies. This initiative can be transformative.

Authors: Risto, Nokia is championing global adoption of 5G, which is going to dramatically accelerate digital engagement. What are some of the safeguards? What are some of the considerations that are being discussed at Nokia about how this accelerating technology can be used for the most good?

Risto: Well, 5G is completely different from the previous generations in a couple of ways. First of all, it really is a holistic overhaul of the digital networking infrastructure we have built over the last 30 years. It can deliver so much more compared to previous generations that our core networks, our fixed-line networks cannot cope with it without further investments. And it is, of course, a much more robust wireless technology.

It can provide much faster data transmission speeds, much higher reliability, much lower latencies, which is necessary when you do something where reaction times need to be extremely quick, like telesurgery. You could perform telesurgery over a wireless connection with 5G. You couldn't do that with 4G . . . or running remotely controlled robots in a hectic environment such as a factory floor.

It's important for humans, consumers, if we start to use much more AR and less VR. Because there you need the same reaction times that the human body has when you turn your head and your eyes see something they didn't see earlier. If you have goggles on and you turn your head, you need to get the new information at the same speed; otherwise you become nauseous. There, also, the speed of light is a constraint.

The data center that sends that information has to be close enough so the speed of light doesn't create delays that would be too long for the human mind, causing us to feel seasick. So 5G is important for the nation-states because it is the backbone of our critical infrastructure in a way that hasn't been used before. It's important for industries, industrial automation, IoT for both consumers and corporates, and then anything that is so heavy-duty stuff that the former networks have not been able to do it, such as AR, VR, telesurgery, industrial automation, and so forth.

So 5G can really do what I talked about earlier in the Future X Network, sort of this Solow's Paradox. 5G is a key part of that solution, but 5G is just one technology. It goes hand in hand with, for example, AI because some of the challenges we face as we push the laws of physics toward the very edge are so complicated that old technologies are not enough to manage them, and therefore we need to use the latest machine-learning technologies to run these networks.

Authors: Elon Musk popularized the thinking of many individuals when he stated, "Mark my words, AI is far more dangerous than nukes," suggesting that a significant portion of the world was going to be displaced by machine learning and artificial intelligence. Of course, he's come at least halfway around the circle on that matter, but it was a valuable sound bite, and it certainly called the issue to public attention.

As we are now accelerating machine learning and the application of AI, we do need to rethink how individuals can perform and/or realize more from the adoption, interaction, and alignment with AI. We were able to participate in a discussion two years ago at Davos with about fifty business leaders where the subject was the workforce of the future.

Less than 10 percent of the people in the room had a five-year plan for what their workforce would look like and how they would have to adjust. Yet, in the same conversation, the moderator asked for input on increases or reduction of workforce that were anticipated. The resulting estimate was a reduction by over 20 percent of the headcount of her companies represented in the room. And we think, if nothing else, it was an eye-opening conversation starter. How are you thinking about that at Nokia? You're one of the most innovative companies in the world. Are your employees worrying about what their role is going to be going forward, and how are you talking about it?

Risto: Well, as with any topic that is such an emotional, let's say catchy subject, of course people are worried about the future and their own future. That's not limited to any single company, but I think that instead of machine learning speeding up, I think it's slowing down, temporarily. There are sort of two dimensions. One dimension is applying the current machine-learning technology broadly, and that we need to do.

There are many companies that don't even understand machine learning at any level. They have not done anything. They are not experimenting yet, and they need to get into this new technology. This is as if electricity would have been invented; you just need to start using it. And you use it to automate repetitive tasks. Basically, for a long time we humans have been doing work that robots should have been doing all along. Now, in a way, we are just giving the robots the work they should have been doing all along.

In this other dimension, which is expanding the areas of applications where machine learning can reduce work in a significant way, I think we are slowing down. We need to come up with some new science before we can tackle meaningful new areas of human work. All the discussion that we can automate 30 percent of the CEO's work—perhaps if the CEO spends one-third of his time doing simple, repetitive tasks, but not if he or she spends time wisely.

But there are societal issues, even if we just apply it in areas where the current technology already is usable. And there is less work to be done by humans, but that work is the simplest, lowest contribution work that we do currently, so the question is: "Can we give these people who are freed up something more valuable to do?" And that's, of course, a challenge.

Thinking about the societal platforms, if we have a platform that encourages experimentation, that encourages people to try out new ways of becoming better themselves, augmenting themselves with technology . . . Let's take an example. We all know that statistically speaking, half of all doctors are below average in their skills. Who wants to go and be treated by a below-average doctor? No one. Using technology, we can lift those up, make them better doctors. That's a huge opportunity, and the same applies to all other work that is similar in nature. So if we have a societal platform that encourages experimentation and reduces the fear of being a victim of automation, I think those societies will do best.

And that's why I have been a strong advocate of new concepts being trialed, such as basic income. And in Finland we have had a basic income experiment for over a year now. We have had 3,000 unemployed people who were randomly picked, who agreed to join the experiment, and they have a basic income model that replaces the standard social services that they use. This is the direction we should take.

How do we reduce the fear? How do we encourage experimentation? How do we create a nimble, agile society? How do we take into account the key areas where this new technology will cause changes so that our legislation, our regulatory frameworks can be ready for that new world instead of us always lagging way behind and preventing rather than encouraging?

Authors: Thank you. That was insightful guidance. We have one last question. Not only are you chairing Nokia, but you're also directly and indirectly influencing a whole legion of innovation pioneers, and I'm wondering what guidance you can give them to ensure that humanity stays at the center of gravity in the man and machine relationship, which really is the core premise of the transHuman code. How are you transferring that thoughtfulness through to the individuals in the organizations that you're influencing?

Risto: Well, we are trying to advocate reflection. For example, if you have time to read my book, the key objective for me to write that book was to make people reflect on the way they are leading their companies, the way they are leading their teams, and whether they should rethink what they are doing. I'm not offering ready-made solutions. I'm telling what happened to Nokia and how we dealt with it, what worked for us.

I'm not claiming that the same behaviors would work for others, but I am encouraging them to think and not be captives of their own roles, if we have a predetermined view of what our role is based on the title we have and we have seen somebody else behave in a particular way having the same title so we tend to do the same things.

But we should open our minds and elevate ourselves and the discussion and the thinking an abstraction layer higher, thinking about, "What is my real duty? Why am I here? What's my goal? What's the goal of my role?" and then rethink how I should behave based on that understanding. As chairman, for example, I view my role quite differently from what is the textbook answer.

Authors: Risto, this has been an enlightening and informative conversation. Nokia has, of course, been the beneficiary of your transformative efforts. Through your book and contribution to local, national, and global collaborations, many more will learn from your experience. Thank you for being a contributor to the transHuman code movement.

Risto: Thank you for the opportunity. ▲

Conversation With Ruma Bose

Co-founder and managing partner of Humanitarian Ventures, global philanthropic champion, and author of *Mother Teresa, CEO: Unexpected Principles for Practical Leadership*

Authors: The transHuman code was envisioned, first, to create a conversation among the many innovators, the implementers, the users of technology that are enabling such dynamic change across every industry sector, and the way we like to look at it is across all the elements of our life ecosystem. And that became the grounding premise of the transHuman Code Initiative, to encourage interaction influenced by what you're thinking and, perhaps even more importantly, what you're doing.

Ruma, you have enjoyed significant success in business, shared your personal experience and guidance through an inspiring bestselling leadership book, *Mother Teresa, CEO*, and are recognized as a driving force, globally, in the social innovation movement. Your experience and outlook on social progress and the role technology can play in this advancement is always valued.

Ruma: Thank you so much. It is always a privilege to talk to you. The primary focus of my current work is finding innovative ways to end the global refugee crisis. My interest in this work started a few years ago when I was leading the Chobani Foundation and Tent Foundation. I was challenged by Hamdi Ulukaya, the founder of Chobani and one of my mentors, to identify what role businesses could play in ending this refugee crisis and what, specifically, we as a company could do? I have found that, too often, when we see a disaster happening in the world, we tend to feel more helpless than we should, when, in fact, if individuals or corporations focus on finding solutions, we can make far more progress than at first seems possible. The refugee crisis is a classic example.

I have now had the opportunity to study in depth the humanitarian sector and understand when it comes to refugees what the challenges are, where the gaps are, and where we could make a meaningful difference. Where I and the Tent Foundation team initially saw a big opportunity was in commissioning new research. Data is quite limited about the humanitarian sector. We also identified that other companies, while interested in helping, did not know what role they could play. With the determination and drive of our team, we were able to build a significant coalition of companies from America and beyond to mobilize and work together to determine how we could have the biggest impact on ending this crisis. We encouraged companies to make commitments to support refugees by hiring refugees, integrating them into supply chains, investing in refugees and delivering services to them.

The next challenge was to expand investment in helping refugees. For instance, the Soros Economic Development Fund (SEDF) committed to invest up to $500 million in businesses that have a positive impact for refugees and migrants.

As an advisor to SEDF, I was tasked with helping them figure out their investment thesis and strategy for deploying this money. This helped me see new opportunities in an area that has long been aid focused. In particular, I saw the opportunity to apply some of the most exciting new technologies in the world to help refugees.

If we take a step back, despite being in the 21st century, the hard truth is that we currently face the biggest refugee crisis in human history.

In part, the challenge lies in that the humanitarian sector and the new technology sector operate in rigid silos that have essentially kept them in two different worlds. The humanitarian sector fundamentally doesn't understand what the technology opportunity is and how technology can help. They're suspicious of it. And the technology industry is not structured to adapt its innovations to meet the needs of the humanitarian sector.

And even when they do recognize that their technology could help, tech founders and investors are often too busy building their businesses to take on the considerable challenge of doing anything about it. Humanitarian causes are not a core part of their business. So they don't know how to execute on them. They wouldn't know who to call, what to do. These two sectors live in parallel. In my current work at Humanitarian Ventures, our focus is to find the cutting-edge technologies with the greatest potential to help and bring them to the humanitarian sector.

Authors: Ruma, we have so often heard you described by others as an enabler. We think from your own life experience, from your desire to make things happen and your sharing skills, you have made it a priority to relay what you've witnessed and/or envisioned could be possible for a larger audience. Some would refer to that as scalability. You like seeing things scale positively.

Ruma: I also feel like a translator. I think very often the challenge is that when people don't speak the same language, in this case business entrepreneurs and humanitarian organizations, the opportunity to make a big positive difference often gets lost in translation. The two sides don't know how to work together. Yet I have found that when there's somebody who can speak in both sides' languages and facilitate and enable their partnership, that's when one plus one becomes three. Let me give you an example to demonstrate what I'm talking about.

There's an AI company called Dataminr that mines massive amounts of data to be able to explain and predict events. So, for example, when there was a chemical weapons attack in Syria, Dataminr was able to report that sarin gas was used far faster than conventional news sources. That extra time saved lives by allowing evacuation to start faster.

Dataminr's technology has been used for some time by the U.S. Army to manage troops on the ground. Yet it was not being used to inform relief organizations that are helping people on the ground. So we, at Humanitarian Ventures, made an investment in Dataminr, and as a result of that, we were able to broker, if I can use that word, a partnership between Dataminr and Mercy Corps, one of the leading NGOs in the humanitarian sector. We're now talking to other humanitarian organizations to pilot the same technology, but what I can tell you is, as of December this past year, Mercy Corps employees in the Middle East now have access to the Dataminr technology.

What the individual in charge of this program at Mercy Corps has told us is that access to live data from Dataminr will allow them to push the most urgent insights to field teams in real-time which could very likely lead to transformational operational outcomes in terms of staff safety and security and decisions on field-level programmatic implementation modalities and locations. Now, our hope is to enable every humanitarian worker in Libya, in Syria, in Yemen, and indeed elsewhere to be able to be safe as a result of this technology. The humanitarian sector at large is one of the last sectors to enjoy the benefits that innovation and technology have brought elsewhere. In part, that is a reflection of cost. But also, when you're busy saving lives and you're used to doing things in a certain way, you don't necessarily make the time to research the new opportunities in the tech sector.

Drone technology, micro-satellite and telecom companies, if humanitarian organizations knew what they were doing in their core markets, they could apply it to help their sector leap decades ahead of where it is today. Often, refugees are using 21st-century technologies, such as mobile phones, to navigate their journeys. Yet, the humanitarian workers who try to help them are often reliant on old technologies such as fax machines. There is a huge opportunity to modernize the refugee relief sector and improve millions of lives by doing so.

Authors: Ruma, we're thankful to be a witness through you to the plight that a growing percentage of the population is facing. We think it would be beneficial for all of our readers to understand what numbers we're talking about, so how many migrants are there today? How many aid workers? How many organizations, and then what little capital is being directed toward this initiative?

Ruma: The numbers are appalling. As of January 2019, there are an estimated 68 million refugees and displaced people in the world. Nearly one person was displaced every two seconds as a result of conflict or persecution in the past year. Three years ago, that number was 60 million. The total is still growing. As we see increasing water wars and refugee migration due to climate change, the exponential growth will be overwhelming. Governments alone can't solve this problem. We estimate that 25 million are refugees who are fleeing violence in their countries by going to another country, and the remainder are internally displaced within their home country. Most are living in poverty with insecurity and without hope.

Half of the 68 million are under the age of 18. About 22 million of them are under the age of five, and about 5 million women are expecting a baby in the next 9 months. The average time that a family or an individual stays in a refugee camp is 17 years. So we are in the process of creating an entire generation of people and families who are growing up in refugee camps, who do not have access to basic healthcare, who do not have access to education, who are growing up angry, hopeless; in short, everything that you wouldn't want for your own kids.

From a human perspective, as I've been dealing with and involved in this crisis, I've seen the absolute best of humanity and I've seen the worst of humanity. I knew that technology can play a huge role in helping us to manage this in a more efficient way.

Authors: Your professional life trajectory has afforded you the opportunity to see the challenges and opportunities of a traditional capitalist market structure. It's afforded you the opportunity to see what impact individuals and organizations can have on addressing the plight of those less fortunate, both at home and abroad. We think it's also afforded you the opportunity to understand how technology companies are conceived and how they evolve and how they succeed. We think the genius in what you're doing is the creation of a business model that allows you to communicate to each of those constituencies and to engage technology creators, providers, enablers to be able to make a contribution to humanitarian causes that would have otherwise gone unserved. And again, here you are enabling that action for the benefit of the end recipients.

Ruma: Thank you.

Authors: The question of course comes back to scalability. You're only one. How are you gonna do this? What does this network look like, and what would you want to tell those people that are interested in this? And we feel thankful to be a part of the dialogue. We'd like to be supportive at least in the smallest way in terms of helping you reach a larger audience. What do we need to tell everybody?

Ruma: So let me just give you a quick overview of what it is that we're doing. We try to solve two big issues. First, how do we bring the most cutting-edge disruptive technologies being built today into the humanitarian sector. Second, how do we get more institutional investors and more private-sector dollars to get behind bringing these ideas into the humanitarian sector?

We are a rules-based fund. Our primary rule is that if one of the top 30 funds in Silicon Valley have invested in a company, then we can co-invest alongside without having to do any extra due diligence. What we've done is analyze the 2,500 deals that were done by these top funds over the past two years and tried to find clear use cases of where these technologies could have an impact in the humanitarian sector. By humanitarian, we do not limit our investment to only refugees and migrants, it also includes emergency response after natural disasters, such as earthquakes and tsunamis.

We were able to identify a hundred potential use cases from our analysis. We are now approaching each of these companies, and if we can convince the CEO to take part in a pilot, then we ask if we can make a small investment in one of their funding rounds (which tend to be in high demand, making it tough for most funds to get into.) Then we get them to outsource delivering humanitarian impact to us. We manage and execute impact for them, recognizing that often or most often, these high-growth companies don't have the bandwidth or the resources to focus on the impact. So far, those we have approached with a potential pilot have all invited us to invest in them.

We are building a different type of impact investment, in which our financial returns and our impact returns are not correlated. We simply co-invest with the top funds in the world, and we take over the impact function and ensure that it actually delivers. We've partnered with multiple NGOs to help us execute on these vital humanitarian innovations.

So, I'd challenge you if you own a technology business, to look at your company and see if a product or service that you've developed might be deployed to help a vulnerable population. If you're an investment firm, copy our model! Let the people who understand impact and know how to execute on impact do that. If you can support the company in pursuing the potential impact, do that. Third, come and work with us! Be our partner, be our investor, be our supporter. Help us scale what we're doing, and be an advocate for humanitarian impact through technological innovation.

Authors: The core premise of the transHuman code was to initiate a conversation, not believing that we had all the answers, but believing that we could start a dialogue among creators and enablers like you and the users and beneficiaries. Knowing that there's an opportunity to create a code and to communicate the code, do you think it can be helpful? Can it be valuable in what you're doing internally and even in your external engagement? You're helping people understand that human beings in partnership with technology result in one plus one equaling three. We are here as an example, and you have so many other examples of people doing incredible things in other sectors. But you're at the heart of the issue. The transHuman code is helping people understand that we're so much more strong and capable when we work in partnership. Technology alone isn't going to solve the world's problems, and humanity alone isn't going to solve the world's problems, but what you're stating and we are proving is that together we can.

Ruma: Today I am thinking about when my five-year-old son grows up, starts learning about his immediate history, during which this refugee crisis is central, and he looks at me and says, "What did you do to help?" I want to be able to have an answer, and it is that which drives me every single day!

Authors: We are very thankful to have you as an ambassador of the transHuman code.

Ruma: I'm very privileged to be an ambassador. Thank you. Lots of gratitude. ▲

Conversation with Jack Faris

Chairman of the Global Alliance to Prevent Prematurity and Stillbirth, social communications pioneer, and multilateral social innovation champion

Authors: Jack, earlier in your career, you led one of the foremost advertising and communication agencies in the Western U.S., but today your focus has evolved to include dynamic social causes, with the Gates Foundation, the Washington Biotechnology and Biomedical Association, the Global Alliance to Prevent Prematurity and Stillbirth, among others.

The application of digital technologies in the communication industry has increased our awareness level of issues and challenges and opportunities dramatically. What's been your experience in communicating, and engaging, and most importantly actioning citizens in an era where the awareness level of good opportunities, for both human and financial capital, is certainly at its highest level?

Jack: During my advertising agency years, I found that when we did communications for our client Boeing that adopted a high level of what I call "generosity of spirit," and used communications that gave credit rather than claimed credit, we did a lot of good for our client, but we also touched people in important ways. So, in one case for example, we did a communication about extending appreciation to soldiers, sailors, airmen, coast guardsmen, at a time when there wasn't much attention being paid to the importance of serving in the military. And their response was strongly positive. It was no attempt to sell Boeing. It was just a big-minded, big-hearted message from Boeing.

And I've found in other circumstances that that can be effective. When authentic, of course. When I was at the Gates Foundation, there were two major programs. One was the Library Project, putting Internet-connected computers into libraries that serve the poorest parts of our country, where at the time they really weren't present at all. And then the other thing was the early years of the global health efforts of the Foundation.

And it was interesting that there was a lot of skepticism and even cynicism about the Library Project, although it was a wonderful project and did a lot of good things. But I think that it was partly that there was at least some plausibility around it being a benefit to Microsoft as well as to users of the Internet in these libraries. Because it was Microsoft software that was being given away and installed, along with a lot of investment in hardware that was very generous of the Foundation.

The Global Health Program, on the other hand, was almost universally acclaimed as being a wonderful thing. So, I'm interested in how we can do things that allow people to connect with something that transcends, at least to a degree, their sense of self-interest. One of the things we did at the Foundation was try to support the global effort to eradicate polio, which had been going on for decades. Which is still underway. We are hoping, very much, to close that out. The theme of the communications effort that we had was, "What's the best way to celebrate the end of the disease?" with the answer being: Let's go after the next one.

Authors: How have you drawn from decades of the development and implementation of communication expertise to benefit the organizations that you've led? What an amazingly, incredibly cluttered communication environment. You know, a proverbial adage in the advertising industry is "the desire, the need, and the success of cutting through the clutter of content." That's still relevant today. There's so much information. Every one of us can be our own producer, our own publisher of content. So maybe, can we take as an example the Global Alliance to Prevent Prematurity and Stillbirth, how are you breaking through with that message?

Jack: Well, let me talk about that. I'd like to put three assets on the table first. One is the unexpected. Two is humanity. And three is celebrity.

In the case of the unexpected, again in the case of Boeing, it was unusual for a major company to be running TV commercials not promoting themselves, but in fact thanking the more than 10 million men and women in uniform.

We did a program for the University of Washington that helped our constituents appreciate the role of the university and what it does in economic vitality and technology development as well as education. And the key strategy was the unexpected one of doing a partnership with what was and still is typically considered to be a major rival, Washington State University. Especially in the field of sports.

So this campaign was organized around the idea of Cougars and Huskies for our economic future. And among others, we had the president of WSU talking about all the great things the University of Washington was doing—in panel presentations, in radio advertising, and in other ways. And that had a terrific response. By objective and empirical measure, it really helped advance the reputation of the university in a way that was very positive.

Also, because we have a lot of brilliant faculty at universities such as UW, they're sometimes too brilliant to be likable unless you're very careful. But one way of presenting what the university does that we found to be quite effective, and this is the humanity part, was students. Having students themselves speak in their own words about what they're doing, what they're experiencing, what they're learning, projects to which they are contributing, just was wildly popular. Faculty are resistible; students are irresistible. And that illustrates the value of humanity.

Now celebrity, that's where we have an asset, in the case of the Global Alliance to Prevent Prematurity and Stillbirth, which, as you know, has a mission of trying to achieve a dramatic reduction in the rate of premature birth worldwide. And it's not incidental to note that premature birth is now the leading cause of death of children worldwide.

The interesting and important benefit of having a celebrity involved, my daughter, Anna Faris—an accomplished actor who can reach a lot of people with a single Instagram post—is now on the GAPS Board and brings her own personal experience of having had a child nine weeks early, who's by the way doing great. Happy, healthy, beautiful, smart. But not all outcomes are as good as that, and she's quite wonderful about inviting the people she can reach to engage, which literally, in an instant, can be millions, to consider supporting the organization, to being interested in the cause, and in our next step, to become participants in a global research project by contributing medical data.

This is in the case of women who are pregnant. Contributing medical data by way of a smartphone app that enables the pregnant woman to carry her medical records, including her pregnancy medical record, everywhere she goes. And at the same time make that data available, de-identified, and aggregated for this global research project, which will seek out new strategies for preventing preterm birth and stillbirth.

So I think that the right use of celebrity can be one of reaching large numbers of people efficiently. We can do things that otherwise might not be possible. If I put something on Instagram, it might reach five people. But in the absence of the kind of advertising budgets that we had to work with for major corporations, the right celebrity with the right, authentic role in a program can be so impactful. Clearly Anna's situation is genuinely authentic. That, I think, can be a powerful strategy in doing something.

Authors: You've referenced a couple of examples, but how do you envision the future of this alliance's engagement with other developers, implementers, and perhaps even financial supporters of technological solutions?

Jack: First, I think it's important to acknowledge, as you and your colleagues do, that we're really on the frontier of this. Some years ago I read a book about the history of the book, and it pointed out that after the invention of the printing press and the widespread application of this new technology, there was rampant plagiarism and appropriation. And the notion of copyright had yet to be invented. I could steal a whole book by Charles Dickens and publish it under my name in America, and there would be little recourse.

And it took about a hundred years to develop the institutions that we now take for granted. Well, I think we're in a similar situation where there's a need for a lot of institutional inventions, and I think your project is contributing very much to that process.

So, as a starting point, I think that's one thing. A couple of things that I thought about with regard to this topic is the power of strategic philanthropy. When I was at the Gates Foundation, I heard firsthand a story from the leader in children's vaccines. He told me that for years, for decades, the whole international project of vaccinating children had fallen into a backwater. The organizations that worked on this had to compete with one another, and there was no ability to collaborate or cooperate.

When Bill and Linda Gates made a $100 million gift to create the Children's Vaccine Program, he said that changed everything. For the first time they could talk about possibilities that were never before even imaginable. And they could work together in ways that were simply impossible previously. And I'm delighted that, in fact, more and more brilliant, successful entrepreneurs, in looking at their philanthropic opportunities, are taking this kind of approach. In this case, the Gates' put this opportunity forward. In other cases, I think that the right collaboration of NGOs and other partners can come forward to present dynamic opportunities to do something dramatic. In fact, it may not surprise you that we're thinking long term about how we might be able to do that in the case of premature birth.

Because the kind of project that can result in a 30–50 percent reduction in premature birth, thereby saving the lives of millions of babies annually, as well as very negative outcomes, would benefit immensely from a well-organized collaborative effort funded at the kind of levels that some of the great initiatives of the Gates Foundation and others have been making.

So thinking about the role of strategic philanthropy, there is so much generosity of spirit and readiness of philanthropic capital available, that the more imagination and aspirational teamwork that we can bring to bear and join up with those who can be persuaded with the right well-designed proposal, the more we can do great things That would be incumbent in that arena.

We have all read the most recent annual letter from Bill and Melinda Gates, and two pieces struck me. One that struck me in that letter is the way that Bill and Linda spent time in in-depth conversations with young African American inmates incarcerated for various crimes and learning about the life that they have been leading and where they came from. And then engaging directly in the kind of programs that are designed to try to help these sorts of kids have a different pathway.

Melinda recently spoke about her experience in Africa. One of the wonderful things about this person is her readiness to go and talk to the women that she's trying to help in their homes. And she said initially, she thought they would be talking about HIV and protecting themselves from infection. But after they got a chance to be alone, the conversation would turn to their interest in contraception. And it's that kind of readiness to dive in that creates not only strategic but intelligent and insightful philanthropy.

I was recently engaged as an advisor to a start-up, VYRTY, and also completely independently with a different cast of characters in a different start-up, SWVL. And as I got to know about them, I saw in both cases the potential application of what they were doing for the mission of the Global Alliance to Prevent Prematurity and Stillbirth. So let me quickly speak to that.

The Global Alliance to Prevent Prematurity and Stillbirth is grounded in three fundamental truths. One is, as I mentioned previously, premature birth is now the leading cause of death of small children worldwide. I think it is reasonable to say it is a major global health problem, if not the most important global health problem. We care very much about children, and by preventing premature birth in a significant way, we would save millions of lives of babies, and we also would save enormous costs of medical care and the costs of disability for those who survive but with lifelong special needs. So it's a hugely important mission.

A second and a little less obvious point is we have amazing technologies in our society to deal with kids who are born early. As I shared earlier, my own grandson was born nine weeks early, spent a month in the NICU with the most incredibly sophisticated and humane, loving care. And he's now just great. And that's the story of many kids who are born early in the US and other advanced countries. But that kind of care is not available to most newborns who come early.

The notion of prevention, as a cultural matter, doesn't have much visibility. People know that premature birth happens and happens often. About one in ten births in America is premature, which is astonishingly high. It's actually worse than the global average. But we don't tend to think about strategies to prevent prematurity. It feels to most like it's an uncontrolled, unpredictable, unpreventable event.

And in fact, the third thing is that we don't know as much as we need to know in order to do a better job of preventing premature birth and stillbirth. So there's a huge scientific enterprise at the heart of this aspiration. Interestingly, it's focused on an area where there has been a research deficit. We have spent enormous amounts of money in trying to understand the dynamics of cancer, and heart disease, and stroke, and arthritis, and other dread illnesses. We have not invested anything like that in trying to understand pregnancy, because we don't think of it as a disease. But it does deserve a lot more emphasis in terms of understanding its dynamics in ways we currently do not.

And so, the Global Alliance to Prevent Prematurity and Stillbirth in concert with many other players, some of them enormously big and important and complicated like USAID and World Health Organization, and companies like Johnson & Johnson, and organizations like March of Dimes, and with generous support from the Gates Foundation, are working collaboratively to build the kind of research capacity and informational resource database of information and tissue samples that enable research on pregnancy that's never before been possible.

And so, we're collecting data at currently 1,700 data elements per woman, on thousands of women, and we hope to expand that in the near future by orders of magnitude with hundreds of thousands of women. This obviously will generate an enormous amount of data. The role of

VYRTY in this is potentially to, as I mentioned earlier, enable pregnant women, women who are anticipating becoming pregnant, to collect all their medical data, include the history of their pregnancy, and contribute that information anonymously to this global research enterprise.

The role of SWVL is an ingenious technology for making the analysis of very large and complicated data sets much more efficient and rapid. And at a much lower cost, it gives us the ability to make sense of what will be an exponentially expanding database. That's the vision. We're at a very early stage, but I'm enormously enthusiastic about its potential.

Authors: This project reflects the convergence of many human-centric initiatives that you've undertaken at different stages in your career. Most importantly, you've realized what the combination of services and organizations are necessary in order to be able to place this as a top-of-mind initiative and to attract the attention necessary. It's possible that the range of skills you've developed over time were always intended to come together for the greatest purpose of all, to lead social innovation.

Jack: That's very kind of you to say. But you know, I realize that neither VYRTY nor SWVL asked me to become a member of their team because I have any kind of technology prowess. Because I don't. And I do want to be modest about what my actual skills and abilities are, but I have had some experience in doing creative work to try to find ways for organizations to pursue good things.

As it happened, of course, the Global Alliance to Prevent Prematurity and Stillbirth is imbued with a sense of mission to do something really important. Fortunately, SWVL and VYRTY both have an ethos of wanting their technologies to be used for good things, not to just become a successful enterprise. So that's what, I think, I might be a little bit good at doing that attracted them to ask me to become part of their team and made possible my connecting each of them to the GAPPS opportunity.

Authors: As a former PhD student, former professor, former university executive, you've been experiencing firsthand how postsecondary institutions can impact the events of society. So, in this, our era of technological transformation, an increasing number of schools are establishing incubators and accelerators to invest both capital and wisdom alongside their students in the developments and application of new innovations. What's been your experience with the development of technology and businesses through these collaborations?

Jack: I've been closer with the University of Washington than other institutions because I worked there. I describe three stages of development in terms of thinking about commercialization. Level one is a sense that it's essentially antithetical. It's a purist attitude that universities should not participate in working with corporations and that it is inherently soiling to be involved in commercial enterprise.

The second level is a sense of benign tolerance. Oh, some people think this is important, so we should acknowledge that and maybe humor them a bit, but keep it under control and make sure that we are protecting our own interests first and foremost.

Level three is wholehearted embracing of this important dynamic with a recognition that failure to do so will result in a long-term, and even short-term, competitive disadvantage, especially, but not only, in terms of human talent both at the student and faculty level. Increasingly, students and faculty want to be part of seeing their inventions, their discoveries, what they work on be part of moving the world forward. And if they don't see an opportunity to do that at one institution, they'll go to another one.So, I think there's been a lot of movement on that scale and much greater sophistication. I do think that smart institutions are beginning to move toward reducing the expectations for year-term benefit in terms of licensing revenues for inventions, and recognizing that long term, the greatest returns on investment of supporting entrepreneurial activities, likely come in the form of grateful philanthropy.

Unshackling entrepreneurs, making it easier to accumulate the capital necessary to see marked inventions and discoveries enter the marketplace and create a return on investment is, I think, another part of that dynamic. The other thing is, when I was at UW, the dean of the business school at the time and I conceived of a social impact business competition program, which we implemented. And it was a wonderful illustration of how you can join the ingenuity of students and faculty and external advisors' technology development but that are designed to benefit people, all of the world, especially in the neediest parts of the world.

Authors: Jack, thank you. You've engaged, informed, and inspired us. And reminded us, in the spirit of the transHuman code, that strength, ingenuity, and promise lie in collaboration. We greatly appreciate your contribution. ▲

THE TRANSHUMAN CODE MANIFESTO

Today, we must acknowledge that we are either building a future of technological grandeur at the expense of what makes us magnificent, or we are building a future of human grandeur with the help of magnificent technology. The path we collectively choose will determine whether our future is bleak or bright. This is humanity's proclamation for choosing wisely. For the sake of our future, we urge you to embrace the following seven declarations:

1. **Privacy:** Securing the privacy of every human being is paramount to realizing the full potential of our future. Therefore, personal data conveyed over the Internet or stored in devices connected to the Internet is owned and solely governed by the individual.

2. **Consent:** Respecting the authority and autonomy of every human being is paramount to realizing the full potential of our future. Therefore, personal digital data will not be used as research, rationale, enticement, or commodity by any entity or individual, except with the explicit, well-informed, revocable consent of the individual owner of the data.

3. **Identity:** Valuing the identity of every human being is paramount

to realizing the full potential of our future. Therefore, everyone everywhere has the right to be known and validated by the possession of a government-issued digital identity, which can be authenticated and used only by its owner.

4. **Ability:** Advancing human faculties is paramount to realizing the full potential of our future. Therefore, to that end, the secure, approved, and accountable aggregation of personal information and resources to increase our individual abilities is a fundamental objective of technology.

5. **Ethics:** Improving the human condition is paramount to realizing the full potential of our future. Therefore, a universal code of ethics reflecting the highest order of human values will govern the development, implementation, and use of technology.

6. **Good:** Advocating and innovating the greatest good for all humanity is paramount to realizing the full potential of our future. Therefore, technology, no matter how advanced, will never supersede the spiritual purposes or the moral rights and responsibilities of any human being anywhere.

7. **Democracy:** Democratizing human vision, ingenuity, and education is paramount to realizing the full potential of our future. Therefore, technology will remain humanity's greatest collaborator but never represent humanity itself.

THE TRANSHUMAN CODE GLOSSARY

POWERED BY WIKIPEDIA

We've chosen to use Wikipedia to supply the definitions in this glossary because it is a fine example of productive, human-centric collaboration between humanity and technology. Further, what led to Jimmy Wales's creation of Wikipedia serves as a reminder of the inherent power available to humanity when we invite the world to contribute to our innovations. In a 2013 interview with Ted Greenwald for *Wired*, Wales explained the catalyst for the creation of Wikipedia in 2001:

> We had been working on Nupedia (the predecessor company) for nearly two years, but we had only completed something like a couple dozen articles. I wanted to figure out why it was taking so long. So I decided to write an entry about Robert Merton, who

had recently won the Nobel Prize in economics. When I set out to do it, I realized that the editors were going to send my draft to the most prestigious finance professors they could find, and it felt very intimidating. That's when I realized, this isn't going to work; it has to be easier to contribute.[1]

Yes, it's been suggested Wikipedia is not, by its very nature, the most credible source, and we understand the sentiment there. Plato would probably conclude the same thing were he alive today—that knowledge can only come from learned sources. However, da Vinci would disagree. The trouble is, placing constraints on the discovery and dissemination of knowledge is one approach that will keep us from making the most of humanity's full spectrum of resources as we seek to create a better future. Yes, the wide-open source approach to innovation and problem-solving is messy and risky. Yes, it's imperfect. But so was the once-credible conclusion that the world was flat.

We believe it is time for us to trust humanity's full array of resources. History has proven that when we pool our means, we can accomplish more than we thought possible. Today's world is one in which the Platos, the da Vincis, and the Bettys and Toms next door have something to offer. Wikipedia offers a microcosmic example of this world. And so, here we provide you with humanity's definitions of the important technological terms that are referenced in the book.

Algorithm, in mathematics and computer science, is an unambiguous specification of how to solve a class of problems. Algorithms can perform calculations, data processing, and automated reasoning tasks. As an effective method, an algorithm can be expressed within a finite amount of space and time and in a well-defined formal language for calculating a function. Starting from an initial state and initial input (perhaps empty), the instructions describe a computation that, when executed, proceeds through a finite number of well-defined successive states, eventually producing output and terminating at a final ending state. The transition from one state to the next is not necessarily deterministic; some algorithms, known as randomized algorithms, incorporate random input.

Artificial General Intelligence (AGI) is the intelligence of a machine that could successfully perform any intellectual task that a human being can. It is a primary goal of some artificial intelligence research and a common topic in science fiction and future studies. Some researchers refer to artificial general intelligence as "strong AI," "full AI," or as the ability of a machine to perform general intelligent action; others reserve "strong AI" for machines capable of experiencing consciousness. Some references emphasize a distinction between strong AI and applied AI (also called "narrow AI" or "weak AI"): the use of software to study or accomplish specific problem solving or reasoning tasks. Weak AI, in contrast to strong AI, does not attempt to perform the full range of human cognitive abilities.

Artificial Intelligence (AI), sometimes called **machine intelligence**, is intelligence demonstrated by machines, in contrast to the **natural intelligence** displayed by humans and other animals. Computer science defines AI research as the study of "intelligent agents": any device that perceives its environment and takes actions that maximize its chance of successfully achieving its goals. AI can be defined as "a system's ability to correctly interpret external data, to learn from such data, and to use those learnings to achieve specific goals and tasks through flexible adaptation." Colloquially, the term "artificial intelligence" is applied when a machine mimics "cognitive" functions that humans associate with other human minds, such as "learning" and "problem-solving."

Authentication (from Greek: αὐθεντικός authentikos, "real, genuine," from αὐθέντης authentes, "author") is the act of confirming the truth of an attribute of a single piece of data claimed true by an entity. In contrast with identification, which refers to the act of stating or otherwise indicating a claim purportedly attesting to a person or thing's identity, authentication is the process of actually confirming that identity. It might involve confirming the identity of a person by validating their identity documents, verifying the authenticity of a website with a digital certificate, determining the age of an artifact by carbon dating, or ensuring that a product is what its packaging and labeling claim to be. In other words, authentication often involves verifying the validity of at least one form of identification.

Big Data refers to data sets, a collection of information, that are too large or complex for traditional data-processing application software to adequately deal with. Data with many cases (rows) offer greater statistical power, while data with higher complexity (more attributes or columns) may lead to a higher false discovery rate. Big data challenges include capturing data, data storage, data analysis, search, sharing, transfer, visualization, querying, updating, information privacy, and data source. Current usage of the term big data tends to refer to the use of predictive analytics, user behavior analytics, or certain other advanced data analytics methods that extract value from data, and seldom to a particular size of data set. Analysis of data sets can find new correlations to "spot business trends, prevent diseases, combat crime, etc."

Behavioral Targeting is centered around the activity or actions of users and is more easily achieved on web pages. Information from browsing websites can be collected from data mining, which finds patterns in users' search history. Advertisers using this method believe it produces ads that will be more relevant to users, thus leading consumers to be more likely influenced by them. If a consumer was frequently searching for plane ticket prices, the targeting system would recognize this and start showing related adverts across unrelated websites, such as airfare deals on Facebook. Its advantage is that it can target an individual's interests, rather than target groups of people whose interests may vary.

A **Blockchain**, originally block chain, is a growing list of records, called blocks, that are linked using cryptography. Each block contains a cryptographic hash of the previous block, a time stamp, and transaction data (generally represented as a Merkle tree root hash). By design, a blockchain is resistant to modification of the data. It is an open, distributed ledger that can record transactions between two parties efficiently and in a verifiable and permanent way. For use as a distributed ledger, a blockchain is typically managed by a peer-to-peer network collectively adhering to a protocol for internode communication and validating new blocks. Once recorded, the data in any given block cannot be altered retroactively without alteration of all subsequent blocks, which requires consensus of the network majority.

Code, in communications and information processing, is a system of rules to convert information—such as a letter, word, sound, image, or gesture—into another form or representation, sometimes shortened or secret, for communication through a communication channel or storage in a storage medium. An early example is the invention of language, which enabled a person, through speech, to communicate what he or she saw, heard, felt, or thought to others. But speech limits the range of communication to the distance a voice can carry and limits the audience to those present when the speech is uttered. The invention of writing, which converted spoken language into visual symbols, extended the range of communication across space and time.

Cryptocurrency (or crypto currency) is a digital asset designed to work as a medium of exchange that uses strong cryptography to secure financial transactions, control the creation of additional units, and verify the transfer of assets. Cryptocurrencies use decentralized control as opposed to centralized digital currency and central banking systems. The decentralized control of each cryptocurrency works through distributed ledger technology, typically a blockchain, that serves as a public financial transaction database. Bitcoin, first released as open-source software in 2009, is generally considered the first decentralized cryptocurrency. Since the release of bitcoin, over 4,000 altcoins (alternative variants of bitcoin, or other cryptocurrencies) have been created.

Cybersecurity, Computer Security, or Information Technology Security (IT security) is the protection of computer systems from theft or damage to their hardware, software, or electronic data, as well as from disruption or misdirection of the services they provide. The field is growing in importance due to increasing reliance on computer systems, the Internet, and wireless networks such as Bluetooth and Wi-Fi, and due to the growth of "smart" devices, including smartphones, televisions, and the various tiny devices that constitute the Internet of things. Due to its complexity, both in terms of politics and technology, it is also one of the major challenges of the contemporary world.

Decentralization is the process by which the activities of an organization, particularly those regarding planning and decision making, are distributed or delegated away from a central, authoritative location or group. Concepts of decentralization have been applied to group dynamics and management science in private businesses and organizations, political science, law and public administration, economics, money, and technology.

A **Digital Identity** is information on an entity used by computer systems to represent an external agent. That agent may be a person, organization, application, or device. ISO/IEC 24760-1 defines identity as a set of attributes related to an entity. The information contained in a digital identity allows for assessment and authentication of a user interacting with a business system on the web, without the involvement of human operators. Digital identities allow our access to computers and the services they provide to be automated and make it possible for computers to mediate relationships. The term "digital identity" has also come to denote aspects of civil and personal identity that have resulted from the widespread use of identity information to represent people in computer systems.

Distributed Object Communication allows objects to access data and invoke methods on remote objects (objects residing in nonlocal memory space). Invoking a method on a remote object is known as remote method invocation (RMI) or remote invocation, and is the object-oriented programming analog of a remote procedure call (RPC). The widely used approach on how to implement the communication channel is realized by using stubs and skeletons. They are generated objects whose structure and behavior depends on the chosen communication protocol, but in general provide additional functionality that ensures reliable communication over the network.

Encryption, in cryptography, is the process of encoding a message or information in such a way that only authorized parties can access it and those who are not authorized cannot. Encryption does not itself prevent interference but denies the intelligible content to a would-be interceptor. In an encryption scheme, the intended information or message, referred to as plaintext, is encrypted using an encryption algorithm—a cipher—generating ciphertext that can be read only if decrypted. For technical reasons, an encryption scheme usually uses a pseudorandom encryption key generated by an algorithm. It is, in principle, possible to decrypt the message without possessing the key, but, for a well-designed encryption scheme, considerable computational resources and skills are required. An authorized recipient can easily decrypt the message with the key provided by the originator to recipients but not to unauthorized users.

The **Fourth Industrial Revolution (4IR)** is the fourth major industrial era since the initial Industrial Revolution of the 18th century. It is characterized by a fusion of technologies that is blurring the lines between the physical, digital, and biological spheres, collectively referred to as cyber-physical systems. It is marked by emerging technology breakthroughs in a number of fields, including robotics, artificial intelligence, nanotechnology, quantum computing, biotechnology, the Internet of Things, the Industrial Internet of Things (IIoT), decentralized consensus, fifth-generation wireless technologies (5G), additive manufacturing/3D printing, and fully autonomous vehicles. Klaus Schwab, the executive chairman of the World Economic Forum, has associated it with the second machine age in terms of the effects of digitization and artificial intelligence (AI) on the global economy. These technologies are disrupting almost every industry in every country. And the breadth and depth of these changes herald the transformation of entire systems of production, management, and governance.

A **Hyperloop** is a proposed mode of passenger and/or freight transportation, first used to describe an open-source vactrain design released by a joint team from Tesla and SpaceX. Drawing heavily from Robert Goddard's vactrain, a hyperloop is a sealed tube or system of tubes through which a pod may travel free of air resistance or friction conveying people or objects at high speed while being very efficient. Elon Musk's version of the concept, first publicly mentioned in 2012, incorporates reduced-pressure tubes in which pressurized capsules ride on air bearings driven by linear induction motors and axial compressors. The Hyperloop Alpha concept was first published in August 2013, proposing and examining a route running from the Los Angeles region to the San Francisco Bay Area, roughly following the Interstate 5 corridor. The Hyperloop Genesis paper conceived of a hyperloop system that would propel passengers along the 350-mile (560 km) route at a speed of 760 mph (1,200 km/h), allowing for a travel time of 35 minutes, which is considerably faster than current rail or air travel times.

The **Internet of Things (IoT)** is the network of devices such as vehicles and home appliances that contain electronics, software, sensors, actuators, and connectivity that allows these things to connect, interact, and exchange data. The IoT involves extending Internet connectivity beyond standard devices, such as desktops, laptops, smartphones, and tablets, to any range of traditionally dumb or non-Internet-enabled physical devices and everyday objects. Embedded with technology, these devices can communicate and interact over the Internet, and they can be remotely monitored and controlled.

A **Massive Open Online Course (MOOC /muːk/)** is an online course aimed at unlimited participation and open access via the web. In addition to traditional course materials, such as filmed lectures, readings, and problem sets, many MOOCs provide interactive courses with user forums to support community interactions among students, professors, and teaching assistants (TAs), as well as immediate feedback to quick quizzes and assignments. MOOCs are a recent and widely researched development in distance education, first introduced in 2006, and emerged as a popular mode of learning in 2012. Early MOOCs often emphasized open-access features, such as open licensing of content, structure, and learning goals, to promote the reuse and remixing of resources. Some later MOOCs use closed licenses for their course materials while maintaining free access for students.

Nanotechnology ("nanotech") is manipulation of matter on an atomic, molecular, and supramolecular scale. The earliest, widespread description of nanotechnology referred to the particular technological goal of precisely manipulating atoms and molecules for fabrication of macroscale products, also now referred to as molecular nanotechnology. A more generalized description of nanotechnology was subsequently established by the National Nanotechnology Initiative, which defines nanotechnology as the manipulation of matter with at least one dimension sized from 1 to 100 nanometers. Because of the variety of potential applications (including industrial and military), governments have invested billions of dollars in nanotechnology research.

Robotics is an interdisciplinary branch of engineering and science that includes mechanical engineering, electronic engineering, information engineering, computer science, and others. Robotics deals with the design, construction, operation, and use of robots, as well as computer systems for their control, sensory feedback, and information processing. These technologies are used to develop machines that can substitute for humans and replicate human actions. Robots can be used in many situations and for lots of purposes, but today many are used in dangerous environments (including bomb detection and deactivation), manufacturing processes, or where humans cannot survive (e.g. in space). Robots can take on any form, but some are made to resemble humans in appearance. This is said to help in the acceptance of a robot in certain replicative behaviors usually performed by people. Many of today's robots are inspired by nature, contributing to the field of bio-inspired robotics.

A **Smart City** is an urban area that uses different types of electronic data collection sensors to supply information that is used to manage assets and resources efficiently. This includes data collected from citizens, devices, and assets that is processed and analyzed to monitor and manage traffic and transportation systems, power plants, water supply networks, waste management, law enforcement, information systems, schools, libraries, hospitals, and other community services. The smart city concept integrates information and communication technology (ICT), and various physical devices connected to the network (the Internet of Things or IoT) to optimize the efficiency of city operations and services and connect to citizens. Smart city technology allows city officials to interact directly with both community and city infrastructure and to monitor what is happening in the city and how the city is evolving.

Spyware is software that aims to gather information about a person or organization, sometimes without their knowledge, that may send such information to another entity without the consumer's consent, that asserts control over a device without the consumer's knowledge, or it may send such information to another entity with the consumer's consent, through cookies. "Spyware" is mostly classified into four types: adware, system monitors, tracking cookies, and trojans; examples of other notorious types include digital rights management capabilities that "phone home," keyloggers, rootkits, and web beacons. Spyware is mostly used for the purposes of tracking and storing Internet users' movements on the Web and serving up pop-up ads to Internet users. Whenever spyware is used for malicious purposes, its presence is typically hidden from the user and can be difficult to detect.

A **Supercomputer** is a computer with a high level of performance compared to a general-purpose computer. The performance of a supercomputer is commonly measured in floating-point operations per second (FLOPS) instead of million instructions per second (MIPS). Since 2017, there are supercomputers that can perform up to nearly a hundred quadrillion FLOPS. Since November 2017, all of the world's fastest 500 supercomputers run Linux-based operating systems. Additional research is being conducted in China, the United States, the European Union, Taiwan, and Japan to build even faster, more powerful, and more technologically superior exascale supercomputers. Supercomputers play an important role in the field of computational science and are used for a wide range of computationally intensive tasks in various fields, including quantum mechanics, weather forecasting, climate research, oil and gas exploration, molecular modeling, and physical simulations. Throughout their history, they have been essential in the field of cryptanalysis.

ACKNOWLEDGMENTS

THE AUTHORS WANT TO ACKNOWLEDGE THE CONTRIBUTIONS of so many in the process of developing the thesis for and writing *The transHuman Code*. Brent Cole, our "collaborator extraordinaire" who tirelessly supported us both in strategy and execution, words cannot (fittingly) express our gratitude for all that you do. Madeleine Morel, the agent's agent, you are truly the embodiment of anything is possible. We credit your embrace of us and our vision in no small part for this realization. Thanks for believing in us. We hope that we made you proud.

We have always been voracious consumers of the writings of others, but with this project we have found new respect for those who write and publish books while continuing to advance their businesses. Wow, this is hard! And without the understanding and support of Peter Ward and Roger Aguinaldo, our respective business partners, it would not have been possible.

Big thanks to Tim Berners-Lee, founder of the World Wide Web, whose conversation with us about the future of the Internet at Davos in January 2016 was the catalyst for this book. We are here on the path because of you and look forward to the journey ahead.

Danil Kerimi, the World Economic Forum, in fact the world, is a better place with you in it! You have been with us from the beginning of *The transHuman Code* journey. Your embrace of this idea and generosity with your community has been invaluable. Rodrigo Arboleda, while MIT gave you an architecture degree, we are so inspired by what you have built that did not involve bricks and mortar. One Laptop Per Child was a fearless program, bringing innovative tech into the far reaches of the developing world in pursuit of digital equality. Don Tapscott, whose best-selling book *The Digital Economy*, written in 1994 and still standing as the definitive pioneer's guide to the future of IT, inspired us both. We are honored to call Don a friend, a colleague, and continue to be guided by his vision and passionate embrace of the future. Wang Wei, we cannot imagine having a better guide to China now and beyond. The future of our countries and their impact on the interface of technology and humanity cannot be understated. Thank you for being our guide.

To the Greenleaf team: Justin Branch, Tyler LeBleu, Lindsey Clark, Justin Parker, Kimberly Lance, Steve Elizalde, and Corrin Foster, you have demonstrated the patience and compassion of wise consummate professionals. We are eternally grateful for your embrace and careful direction.

Lastly, we are thankful to our families, who have continuously supported our wanderlust for creating what can be. Only our wives, Tracy (David) and Anne (Carlos), knew, of course, that this adventure to foresee the future would consume us beyond reason. Thank you for continuing to allow us to dream and play!

NOTES

Introduction: How to Read This Book

1. Luc de Clapiers, Marquis of Vauvenargües, *Reflections and Maxims of Vauvenargues*: Translated into English by R. G. Stevens (London: Humphrey Milford, 1940), 186–87.
2. "26 Astounding Facts about the Human Body," MSN.com, April 28, 2018, https:// www.msn.com/en-in/health/medical/26-astounding-facts-about-the-human-body /ss-BBl9CfX#image=19; AND https://www.independent.co.uk/life-style/health -and-families/features/18-facts-you-didnt-know-about-how-amazing-your-body -is-a6725486.html; VIA https://www.youtube.com/watch?v=tozEuziqdpg.
3. Deepak Chopra, *Quantum Healing* (New York: Bantam, 1989), 262–63, https://itunes.apple.com/us/book/quantum-healing-revised-and-updated /id1013581458?mt=11.
4. Dr. Werner Gitt, "Information: The Third Fundamental Quantity," *Siemens Review* 56 (November/December 1989), 2–7.

Chapter 1: The Pinnacle and Purpose of Technology

1. Kevin Kelly, *The Inevitable* (New York: Penguin, 2016), 18, https://itunes .apple.4com/us/book/the-inevitable/id1048849451?mt=11.
2. Erik Weiner, "The Cost of Saying Yes to Convenience," *LA Times*, June 1, 2015.
3. Ryan Whitwam, "IBM, Department of Energy Unveil World's Fastest Supercomputer," ExtremeTech, June 8, 2018, https://www.extremetech .com/extreme/271005-ibm-department-of-energy-unveil-summit-the-worlds -fastest-supercomputer?utm_source=email&utm_campaign=extremetech&utm _medium=title.
4. Kelly, *The Inevitable*.

5. Kelly, *The Inevitable*.

6. Vernor Vinge, "The Coming Technological Singularity: How to Survive in the Post-Human Era," written for the VISION-21 Symposium sponsored by NASA Lewis Research Center and the Ohio Aerospace Institute, March 30–31, 1993.

7. Elon Musk via @Elonmusk on Twitter at 12:54 PM on April 13, 2018.

8. Susan Fowler, "'What Have We Done?': Silicon Valley Engineers Fear They've Created a Monster," *Vanity Fair*, August 9, 2018, https://www.vanityfair.com /news/2018/08/silicon-valley-engineers-fear-they-created-a-monster.

9. Stuart Jeffries, "How the Web Lost Its Way – And Its Founding Principles," *The Guardian*, August 24, 2014, https://www.theguardian.com/technology/2014 /aug/24/internet-lost-its-way-tim-berners-lee-world-wide-web.

10. Ibid.

11. Ibid.

12. Rebecca Walker Reczek, Christopher Summers, Robert Smith, "Targeted Ads Don't Just Make You More Likely to Buy—They Can Change How You Think About Yourself," April 4, 2016, https://hbr.org/2016/04/targeted-ads-dont-just-make -you-more-likely-to-buy-they-can-change-how-you-think-about-yourself.

13. Ibid.

14. Ibid.

15. Katrina Brooker, "'I Was Devastated': Tim Berners-Lee, the Man Who Created the World Wide Web, Has Some Regrets," *Vanity Fair*, July 1, 2018, https://www .vanityfair.com/news/2018/07/the-man-who-created-the-world-wide-web-has -some-regrets.

16. Klaus Schwab, *The Fourth Industrial Revolution* (Geneva, Switzerland: World Economic Forum, 2016), 15, https://itunes.apple.com/us/book/the-fourth -industrial-revolution/id1139621463?mt=11.

Chapter 2: Framing Our Best Future

1. "Who Is Abraham Maslow and What Are His Contributions to Psychology," PositivePsychologyProgram, September 29, 2017, https:// positivepsychologyprogram.com/abraham-maslow/#needs-abrahammaslow.

2. Abraham Maslow, "A Theory of Human Motivation," *Psychological Review* 50, no. 4 (1943), 370–396, http://dx.doi.org/10.1037/h0054346.

3. "Who Is Abraham Maslow and What Are His Contributions to Psychology," PositivePsychologyProgram, September 29, 2017, https:// positivepsychologyprogram.com/abraham-maslow/#needs-abrahammaslow.

4. Abraham Maslow, "A Theory of Human Motivation," *Psychological Review* 50, no. 4 (1943), 370–396, http://dx.doi.org/10.1037/h0054346.

5. "Who Is Abraham Maslow and What Are His Contributions to Psychology," PositivePsychologyProgram, September 29, 2017, https://positivepsychologyprogram.com/abraham-maslow/#needs-abrahammaslow.

6. Vinod Khosla, "Reinventing Societal Infrastructure with Technology," Medium.com, accessed December 20, 2018, https://medium.com/@vkhosla/reinventing-societal-infrastructure-withtechnology-f71e0d4f2355.

Chapter 3: Water

1. Scott Harrison & Lisa Sweetingham, *Thirst* (New York: Currency, 2018), 611–615, https://itunes.apple.com/us/book/thirst/id1323714868?mt=11.

2. Ibid.

3. Charity: Water website, accessed December 20, 2018, https://www.charitywater.org/.

4. "Global Water, Sanitation, & Hygiene (WASH)," CDC, accessed December 20, 2018, https://www.cdc.gov/healthywater/global/wash_statistics.html.

5. "Amazon Water Comprehensively Mapped from Space," ScienceDaily.com, June 24, 2014, https://www.sciencedaily.com/releases/2014/06/140624093236.htm.

6. "Stanford Breakthrough Provides Picture of Underground Water," Stanford.edu, June 17, 2014, https://news.stanford.edu/pr/2014/pr-radar-groundwater-woods-061614.html.

7. Ibid.

8. Harrison and Sweetingham, *Thirst*, 620.

9. "Waterseer," accessed December 20, 2018, https://www.waterseer.org/ AND Will Henley, "The New Water Technologies That Could Save the Planet," *The Guardian*, July 22, 2013, https://www.theguardian.com/sustainable-business/new-water-technologies-save-planet.

10. Celeste Hicks, "'Cloud Fishing' Reels in Precious Water for Villagers in Rural Morocco," *The Guardian*, December 26, 2016, https://www.theguardian.com/global-development/2016/dec/26/cloud-fishing-reels-in-precious-water-villagers-rural-morocco-dar-si-hmad.

11. Rosie Spinks, "Could These Five Innovations Help Solve the Global Water Crisis?" *The Guardian*, February 13, 2017, https://www.theguardian.com/global-development-professionals-network/2017/feb/13/global-water-crisis-innovation-solution.

12. "Diarrhoeal Disease," WHO, May 2, 2017, http://www.who.int/news-room/fact-sheets/detail/diarrhoeal-disease.

13. Will Henley, "The New Water Technologies That Could Save the Planet," *The Guardian*, July 22, 2013, https://www.theguardian.com/sustainable-business/new-water-technologies-save-planet.

14. "Water," Emory.edu, accessed December 20, 2018, https://sustainability.emory
.edu/initiatives/water/.

15. Robert McMillan, "Hackers Break into Water System Network," ComputerWorld,
October 31, 2006, https://www.computerworld.com/article/2547938/security0
/hackers-break-into-water-system-network.html.

16. Görrel Espelund, "How Vulnerable Are Water Utilities to Traditional and Cyber
Threats?" ESE Magazine, May 9, 2016, https://esemag.com/featured/how
-vulnerable-are-water-utilities-to-cyber-threats/.

17. "The Impact of a Cotton T-Shirt," WWF, January 16, 2013,
https://www.worldwildlife.org/stories/the-impact-of-a-cotton-t-shirt.

Chapter 4: Food

1. From an interview between the authors and Dr. Fraser on April 27, 2018.

2. Ibid.

3. "Global Agriculture Towards 2050," FAO.org, accessed December 20, 2018,
http://www.fao.org/fileadmin/templates/wsfs/docs/Issues_papers
/HLEF2050_Global_Agriculture.pdf.

4. "Food Per Person," Our World in Data, accessed December 20, 2018,
https://ourworldindata.org/food-per-person.

5. "Globally Almost 870 Million Chronically Undernourished - New Hunger
Report," FAO.org, accessed December 20, 2018, http://www.fao.org/news/story
/en/item/161819/icode/.

6. "2 Billion Worldwide Are Obese or Overweight," Consumer HealthDay.com,
accessed December 20, 2018, https://consumer.healthday.com/vitamins-and
-nutrition-information-27/obesityhealth-news-505/2-billion-worldwide-are-obese
-or-overweight-723536.html.

7. "One Third of World's Food Is Wasted, Says UN Study," BBC.com, May 11, 2011,
https://www.bbc.com/news/world-europe-13364178.

8. Fraser interview.

9. Fraser interview.

10. Fraser interview.

11. Bryan Walsh, "The Triple Whopper Environmental Impact of Global Meat
Production," Time.com, December 16, 2013, http://science.time.com/2013/12/16
/the-triple-whopper-environmental-impact-of-global-meat-production/.

12. "Food Per Person," Our World in Data, accessed December 20, 2018,
https://ourworldindata.org/food-per-person.

13. Bryan Walsh, "The Triple Whopper Environmental Impact of Global Meat
Production," Time.com, December 16, 2013, http://science.time.com/2013/12/16
/the-triple-whopper-environmental-impact-of-global-meat-production/.

14. Jess McNally, "Can Vegetarianism Save the World? Nitty-gritty," Stanford.edu, August 31, 2011, https://alumni.stanford.edu/get/page/magazine/article/?article _id=29892.

15. Bryan Walsh, "The Triple Whopper Environmental Impact of Global Meat Production," Time.com, December 16, 2013, http://science.time.com/2013/12/16 /the-triple-whopper-environmental-impact-of-global-meat-production/.

16. "Livestock's Long Shadow," FAO.org, accessed December 20, 2018, 112, http://www.fao.org/docrep/010/a0701e/a0701e.pdf.

17. Christina Troitino, "Memphis Meats' Lab-Grown Meat Raises $17M with Help from Bill Gates and Richard Branson," Forbes.com, August 24, 2017, https://www .forbes.com/sites/christinatroitino/2017/08/24/memphis-meats-lab-grown-meat -raises-17m-with-help-from-bill-gates-and-richard-branson/#3c92b5613fd0.

18. Fraser interview.

19. Fraser interview.

Chapter 5: Security

1. Drew Armstrong, "My Three Years in Identity Theft Hell," Bloomberg.com, September 13, 2017, https://www.bloomberg.com/news/articles/2017-09-13/my -three-years-in-identity-theft-hell.

2. Jeff John Roberts, "Home Depot to Pay Banks $25 Million in Data Breach Settlement," Fortune.com, March 9, 2017, http://fortune.com/2017/03/09/home -depot-data-breach-banks/.

3. Nick Wells, "How the Yahoo Hack Stacks Up to Previous Data Breaches," CNBC .com, October 4, 2017, https://www.cnbc.com/2017/10/04/how-the-yahoo-hack -stacks-up-to-previous-data-breaches.html.

4. Sissi Cao, "The 5 Most Notable Cybersecurity Breaches—and Aftermath," Observer.com, November 29, 2017, http://observer.com/2017/11/the-5-most -notable-cybersecurity-breaches-andaftermath/.

5. Ibid.

6. "17.6 Million U.S. Residents Experienced Identity Theft in 2014," BJS.gov, September 27, 2015, https://www.bjs.gov/content/pub/press/vit14pr.cfm.

7. Dan Patterson, "Why Trust Is the Essential Currency of Cybersecurity," TechRepublic.com, May 23, 2018, https://www.techrepublic.com/article/why -trust-is-the-essential-currency-of-cybersecurity/.

8. Ibid.

9. Ibid.

10. Tom Simonite, "Tech Firms Move to Put Ethical Guard Rails Around AI," WIRED.com, May 16, 2018, https://www.wired.com/story/tech-firms-move-to-put-ethical-guard-rails-around-ai/.

11. Ibid.

12. Ibid.

Chapter 6: Health

1. "Tithonus," Britannica.com, accessed December 20, 2018, https://www.britannica.com/topic/Tithonus-Greek-mythology.

2. Ibid.

3. M. Nathaniel Mead, "Nutrigenomics: The Genome–Food Interface," *Environmental Health Perspectives* 115, no. 12 (December 2007), A582–A589, https://www.ncbi.nlm.nih.gov/pmc/articles/PMC2137135/.

4. Ibid.

5. Ibid.

6. P. Tricoci, JM Allen, et al., "Scientific Evidence Underlying the ACC/AHA Clinical Practice Guidelines," *JAMA* 301, no. 8 (February 25, 2009), 831–41, https://www.ncbi.nlm.nih.gov/pubmed/19244190.

7. AliveCor, accessed December 20, 2018, https://www.alivecor.com/.

8. "Transforming Patient Care with the Power of AI," Zebra, accessed December 20, 2018, https://www.zebra-med.com/.

9. TwoPoreGuys, accessed December 20, 2018, https://twoporeguys.com/.

10. Neurotrack, accessed December 20, 2018, https://www.neurotrack.com/.

Chapter 7: Jobs

1. Susan Fowler, "'What Have We Done?': Silicon Valley Engineers Fear They've Created a Monster," VanityFair.com, August 9, 2018, https://www.vanityfair.com/news/2018/08/silicon-valley-engineers-fear-they-created-a-monster.

2. Catey Hill, "10 Jobs Robots Already Do Better Than You," MarketWatch.com, January 27, 2014, https://www.marketwatch.com/story/9-jobs-robots-already-do-better-than-you-2014-01-27.

3. Ibid.

4. Ibid.

5. Ibid.

6. Russell Heimlich, "Baby Boomers Retire," Pew Research.org, December 29, 2010, http://www.pewresearch.org/fact-tank/2010/12/29/baby-boomers-retire/.

7. Catey Hill, "10 Jobs Robots Already Do Better Than You," MarketWatch.com, January 27, 2014, https://www.marketwatch.com/story/9-jobs-robots-already-do-better-than-you-2014-01-27.

8. Al Gini, *The Importance of Being Lazy* (New York: Routledge, 2003), 32.

9. Todd Duncan, *Time Traps* (Nashville: Thomas Nelson, 2006), 189.

Chapter 8: Money

1. From a conversation at the World Economic Forum between the authors and Don Tapscott on January 23, 2018.

2. Tapscott conversation.

3. "Bitcoin: A Peer-to-Peer Electronic Cash System," Bitcoin Wiki Essays, accessed December 20, 2018, https://en.bitcoin.it/wiki/Essay:Bitcoin:_A_Peer-to-Peer_Electronic_Cash_System.

4. Ibid.

5. Ibid.

Chapter 9: Transportation

1. Dara Kerr, "Electric Scooters Are Invading. Bird's CEO Leads the Charge," CNET.com, April 24, 2018, https://www.cnet.com/news/the-electric-scooter-invasion-is-underway-bird-ceo-travis-vanderzanden-leads-the-charge/.

2. "People In San Francisco Are Pissed Over These Electric Scooters," May 2, 2018, https://www.youtube.com/watch?v=T2SK_60VpHs.

3. "Bird Rides, Inc. Agrees to Plead 'No Contest' in Violating City Law and Will Pay Over $300,000 in Fines and Restitution," Santa Monica.gov, May 2, 2018, https://www.santamonica.gov/birdpleaagreement.

4. Dara Kerr, "Electric Scooters Are Invading. Bird's CEO Leads the Charge," CNET.com, April 24, 2018, https://www.cnet.com/news/the-electric-scooter-invasion-is-underway-bird-ceo-travis-vanderzanden-leads-the-charge/.

5. Donald Wood, "When Will Rolls-Royce Introduce Flying Taxis?" TravelPulse.com, July 17, 2018, https://www.travelpulse.com/news/travel-technology/when-will-rolls-royce-introduce-flying-taxis.html.

6. Melissa Locker, "Inside Zunum Aero's hybrid-electric plane," FastCompany.com, August 22, 2018, https://www.fastcompany.com/90211809/inside-zunum-aeros-hybrid-electricplane.

7. Marina Koren, "What Would Flying From New York to Shanghai in 39 Minutes Feel Like?" The Atlantic.com, October 3, 2017, https://www.theatlantic.com/technology/archive/2017/10/spacex-elon-musk-mars-moon-falcon/541566/.

8. Vinod Khosla, "Reinventing Societal Infrastructure with Technology," Medium
   .com, accessed December 20, 2018, https://medium.com/@vkhosla/reinventing
   -societal-infrastructure-withtechnology-f71e0d4f2355.

9. Shafi Musaddique, "Here Are the World's 10 Most Polluted Cities – 9 Are in India,"
   CNBC.com, May 3, 2018, https://www.cnbc.com/2018/05/03/here-are-the
   -worlds-10-most-polluted-cities--9-are-in-india.html.

10. "Number of International Tourist Arrivals Worldwide from 1996 to 2017 (in
   Millions)," Statista.com, accessed December 20, 2018, https://www.statista.com
   /statistics/209334/total-number-of-international-touristarrivals/.

Chapter 10: Communication

1. Colin Lecher, "French Presidential Candidate Mélenchon Uses 'Hologram' Optical
   Illusion to Appear in Seven Places," The Verge.com, April 19, 2017, https://www
   .theverge.com/2017/4/19/15357360/melenchon-france-electionhologram.

2. "France Election: Hard-Left Candidate Melenchon Appears by Hologram," BBC
   .com, February 5, 2017, https://www.bbc.com/news/av/world-europe-38875197
   /france-election-hard-left-candidate-melenchon-appears-by-hologram.

3. Rachel Donadio, "A French Campaign Waged Online Adds a Wild Card to the
   Election," NYTimes.com, April 22, 2017, https://www.nytimes.com/2017/04/22
   /world/europe/france-election-jean-luc-melenchon-web.html.

4. Amy Mitchell, Heather Brown, and Emily Guskin, "The Role of Social Media in the
   Arab Uprisings," Journalism.org, November 28, 2012, http://www.journalism
   .org/2012/11/28/role-social-media-arab-uprisings/.

5. Simon Kemp, "Digital in 2018: World's Internet Users Pass the 4 Billion Mark,"
   WeAreSocial.com, January 30, 2018, https://wearesocial.com/uk/blog/2018/01
   /global-digital-report-2018.

6. Shalina Misra et al., "The iPhone Effect: The Quality of In-Person Social
   Interactions in the Presence of Mobile Devices," *Environment and Behavior* 48, no. 2
   (2016), http://journals.sagepub.com/doi/abs/10.1177/0013916514539755.

7. Emily Drago, "The Effect of Technology on Face-to-Face Communication,"
   Elon.edu, accessed December 20, 2018, https://www.elon
   .edu/u/academics/communications/journal/wp-content/uploads
   /sites/153/2017/06/02DragoEJSpring15.pdf.

8. The interaction took place at the 2017 M&A Advisor Summit in New York City on
   November 13, 2017.

9. M&A Advisor Summit, November 2017.

10. P. J. Manney, "Is Technology Destroying Empathy?" LiveScience.com, June 30, 2015, https://www.livescience.com/51392-will-tech-bring-humanity-together-or -tear-it-apart.html.

11. Aaron Blake, "A New Study Suggests Fake News Might Have Won Donald Trump the 2016 Election," WashingtonPost.com, April 3, 2018, https://www .washingtonpost.com/news/the-fix/wp/2018/04/03/a-new-study-suggests-fake -news-might-have-won-donald-trump-the-2016-election/?noredirect=on&utm _term=.ad4325386328.

Chapter 11: Community

1. Danilo Matoso Macedo and Sylvia Ficher "Brasilia: Preservation of a Modernist City," Getty.edu, accessed December 20, 2018, http://www.getty.edu /conservation/publications_resources/newsletters/28_1/brasilia.html.

2. Christopher Hawthorne, "Critic's Notebook: Brasilia's Embrace of the Future Seems So Quaint," LATimes.com, April 21, 2010, http://articles.latimes.com/2010 /apr/21/entertainment/la-et-brasilia-20100421.

3. Ana Nicolaci Da Costa, "50 Years On, Brazil's Utopian Capital Faces Reality," Reuters.com, April 21, 2010, https://www.reuters.com/article/us-brazil-brasilia/50 -years-on-brazils-utopian-capital-faces-reality-idUSTRE63K4CT20100421.

4. Okulicz-Kozaryn Adam: When Place is Too Big: Happy Town and Unhappy Metropolis, 55th Congress of the European Regional Science Association: "World Renaissance: Changing roles for people and places," August 25–28, 2015, Lisbon, Portugal, European Regional Science Association (ERSA), Louvain-la-Neuve, https://www.econstor.eu/bitstream/10419/124581/1/ERSA2015_00148.pdf.

5. Barbara Vobejda, "Legacy of Urban Sprawl: Desolation and Isolation," WashingtonPost.com, February 12, 1993, https://www.washingtonpost.com /archive/politics/1993/02/12/legacy-of-urban-sprawl-desolation-and -isolation/4fb940af-a88c-439a-b7cd-a03db70a58a4/?utm_term=.876d1d2bf04d.

6. Icon Build, accessed December 20, 2018, https://www.iconbuild.com/home.

Chapter 12: Education

1. "The Social History of the MP3," Pitchfork.com, accessed December 20, 2018, https://pitchfork.com/features/article/7689-the-social-history-of-the-mp3/?page=2.

2. "The mp3 History," mp3-History.com, accessed December 20, 2018, https://www.mp3-history.com/en/timeline.html.

3. Mark Sweney, "Slipping Discs: Music Streaming Revenues of $6.6bn Surpass CD Sales," TheGuardian.com, April 24, 2018, https://www.theguardian.com /technology/2018/apr/24/music-streaming-revenues-overtake-cds-to-hit-66bn.

4. Robyn D. Shulman, "EdTech Investments Rise to a Historical $9.5 Billion: What Your Startup Needs to Know," *Forbes*, January 26, 2018, https://www.forbes.com/sites/robynshulman/2018/01/26/edtech-investments-rise-to-a-historical-9-5-billion-what-your-startup-needs-to-know/#63afe9533a38.

5. David Raths, "edX CEO: 'It Is Pathetic That the Education System Has Not Changed in Hundreds of Years,'" CampusTechnology.com, July 31, 2014, https://campustechnology.com/Articles/2014/07/31/edX-CEO-It-Is-Pathetic-That-the-Education-System-Has-Not-Changed-in-Hundreds-of-Years.aspx?Page=1.

6. "Teachers' Dream Classroom," EdTechRoundup.org, March 31, 2016, http://www.edtechroundup.org/uploads/2/6/5/7/2657242/edgenuity_dream_classroom_report_033116_final.pdf.

7. Ashley Southall, "Charles M. Vest, 72, President of M.I.T. and a Leader in Online Education, Dies," NYTimes.com, December 16, 2013, https://www.nytimes.com/2013/12/16/us/charles-m-vest-72-president-of-mit-and-a-leader-in-online-education-dies.html?_r=0.

8. Tamar Lewin, "The Evolution of Higher Education," NYTimes.com, November 6, 2011, https://www.nytimes.com/2011/11/06/education/edlife/the-evolution-of-higher-education.html.

9. David Price, *Open* (Great Britain: Crux Publishing, 2013), 280–81, https://itunes.apple.com/us/book/open/id871218678?mt=11.

10. "Vinod Khosla," Crunchbase.com, accessed December 20, 2018, https://www.crunchbase.com/person/vinod-khosla.

11. Buckminster Fuller, *Critical Path* (New York: St. Martins Press, 1981).

Chapter 13: Government

1. Carole Cadwalladr, "'I Made Steve Bannon's Psychological Warfare Tool': Meet the Data War Whistleblower," TheGuardian.com, March 17, 2018, https://www.theguardian.com/news/2018/mar/17/data-war-whistleblower-christopher-wylie-faceook-nix-bannon-trump.

2. Christopher Wylie, "Christopher Wylie: Why I Broke the Facebook Data Story – And What Should Happen Now," TheGuardian.com, April 7, 2018, https://www.theguardian.com/uk-news/2018/apr/07/christopher-wylie-why-i-broke-the-facebook-data-story-and-what-should-happen-now.

3. Carole Cadwalladr, "'I Made Steve Bannon's Psychological Warfare Tool': Meet the Data War Whistleblower," TheGuardian.com, March 17, 2018, https://www.theguardian.com/news/2018/mar/17/data-war-whistleblower-christopher-wylie-faceook-nix-bannon-trump.

4. Christopher Wylie, "Christopher Wylie: Why I Broke the Facebook Data Story – And What Should Happen Now," TheGuardian.com, April 7, 2018, https://www.theguardian.com/uk-news/2018/apr/07/christopher-wylie-why-i-broke-the-facebook-data-story-and-what-should-happen-now.

5. "Full Wylie Interview: 'Very difficult to verify' whether Facebook data has been purged," NBCnews.com, accessed December 20, 2018, https://www.nbcnews.com/meet-the-press/video/full-wylie-interview-very-difficult-to-verify-whether-facebook-data-has-been-purged-1205607491661.

6. Stephen Lam, "Facebook to send Cambridge Analytica data-use notices to 87 million users Monday," NBCnews.com, https://www.nbcnews.com/tech/social-media/facebook-send-cambridge-analytica-data-use-notices-monday-n863811.

7. Ibid.

8. "Cambridge Analytica whistleblower: 'We spent $1m harvesting millions of Facebook profiles,'" YouTube.com, accessed December 20, 2018, https://www.youtube.com/watch?time_continue=394&v=FXdYSQ6nu-M.

9. Edmund L. Andrews, "The Science Behind Cambridge Analytica: Does Psychological Profiling Work?" Stanford.edu, April 12, 2018, https://www.gsb.stanford.edu/insights/science-behind-cambridge-analytica-does-psychological-profiling-work.

10. Kristin Houser, "5 Ways That Technology Is Transforming Politics in the Age of Trump," Futurism.com, February 8, 2017, https://futurism.com/5-ways-that-technology-is-transforming-politics-in-the-age-of-trump/.

11. Andrew James Benson, "Liquid Democracy with Santiago Siri," YouTube.com, April 1, 2017, https://www.youtube.com/watch?v=nBxauY1f36A.

Chapter 14: Innovation

1. Alex Wilson, "Kelly Slater's Wave Pool Is the Future. And It's Bleak," Outside.com, May 7, 2018, https://www.outsideonline.com/2303871/world-surf-league-founders-cup.

2. Ibid.

3. "Jack O'Neill, Surfer Who Made the Wetsuit Famous, Dies at 94," NYTimes.com, June 5, 2017, https://www.nytimes.com/2017/06/05/business/jack-oneill-dead-popularized-the-wet-suit.html.

4. Alex Wilson, "Obituary: Jack O'Neill (1923–2017)," Outside.com, June 4, 2017, https://www.outsideonline.com/2190246/obituary-jack-oneill-1923-2017.

5. "Jack O'Neill, Surfer Who Made the Wetsuit Famous, Dies at 94," NYTimes.com, June 5, 2017, https://www.nytimes.com/2017/06/05/business/jack-oneill-dead-popularized-the-wet-suit.html.

6. Alex Wilson, "Obituary: Jack O'Neill (1923–2017)," Outside.com, June 4, 2017, https://www.outsideonline.com/2190246/obituary-jack-oneill-1923-2017.

7. Andrew James Benson, "Liquid Democracy with Santiago Siri," YouTube.com, April 1, 2017, https://www.youtube.com/watch?v=nBxauY1f36A.

8. *Soul Surfer*, directed by Sean McNamara and distributed by FilmDistrict and TriStar Pictures (2011). The film is based on the 2004 autobiography *Soul Surfer: A True Story of Faith, Family, and Fighting to Get Back on the Board* by Bethany Hamilton.

9. Chris Burkard, in an interview with Victoria Sambursky for digitaltrends.com, https://www.digitaltrends.com/outdoors/chris-burkard-interview-under-an-arctic -sky/.

10. Burkard interview.

11. Burkard interview.

12. "Peary's expedition reaches North Pole?" History.com, accessed December 20, 2018, https://www.history.com/this-day-in-history/pearys-expedition-reaches -north-pole.

13. Our thanks to author Erik Wahl for bringing this beautiful story to our attention in both personal conversation and in his book *The Spark and the Grind*.

14. Norman Ollestad, *Crazy for the Storm* (New York: Harper Collins, 2009), 101–20.

15. Brené Brown, *Rising Strong* (New York: Spiegel & Grau/Random House, 2015), 7.

The transHuman Code Glossary

1. Ted Greenwald, "How Jimmy Wales' Wikipedia Harnessed the Web as a Force for Good," *Wired*, March 19, 2013, https://www.wired.com/2013/03/jimmy-wales- wikipedia/.

ABOUT THE AUTHORS

CARLOS MOREIRA IS A MULTI–AWARD WINNING TECHNOLOGY PIONEER whose IT, online security, and trust management experience with the United Nations and World Trade Organization guided the creation, twenty years ago, of one of the world's first cybersecurity companies, WISeKey, which he leads today. He is also an active leader and member of several institutions and organizations focused on the advancement of technological innovation and preservation of human identity. He resides in Geneva, Switzerland with his family.

DAVID FERGUSSON IS A CORPORATE FINANCE LEADER SPECIALIZING in global mergers and acquisitions. He engages regularly with business, media, political, and academic leaders on the factors influencing corporate growth. He is a pioneer and international award winner for cross-border investment between the United States and China. Executive Director of M&A for Generational Equity and Chairman of the finance industry's leading think tank, he is recognized as an expert on the impact of technology on business, government, and humankind. He resides with his family in Westchester, New York.

INDEX